Advanced Structural Analysis with MATLAB®

Advanced Structural Analysis with MATLAB®

By
Srinivasan Chandrasekaran

CRC Press
Taylor & Francis Group
Boca Raton London New York

CRC Press is an imprint of the
Taylor & Francis Group, an **informa** business

MATLAB® and Simulink® are trademarks of the MathWorks, Inc. and are used with permission. The MathWorks does not warrant the accuracy of the text or exercises in this book. This book's use or discussion of MATLAB® and Simulink® software or related products does not constitute endorsement or sponsorship by the MathWorks of a particular pedagogical approach or particular use of the MATLAB® and Simulink® software.

CRC Press
Taylor & Francis Group
6000 Broken Sound Parkway NW, Suite 300
Boca Raton, FL 33487-2742

© 2019 by Taylor & Francis Group, LLC
CRC Press is an imprint of Taylor & Francis Group, an Informa business

No claim to original U.S. Government works

Printed on acid-free paper

International Standard Book Number-13: 978-0-367-02645-5 (Hardback)

This book contains information obtained from authentic and highly regarded sources. Reasonable efforts have been made to publish reliable data and information, but the author and publisher cannot assume responsibility for the validity of all materials or the consequences of their use. The authors and publishers have attempted to trace the copyright holders of all material reproduced in this publication and apologize to copyright holders if permission to publish in this form has not been obtained. If any copyright material has not been acknowledged, please write and let us know so we may rectify in any future reprint.

Except as permitted under U.S. Copyright Law, no part of this book may be reprinted, reproduced, transmitted, or utilized in any form by any electronic, mechanical, or other means, now known or hereafter invented, including photocopying, microfilming, and recording, or in any information storage or retrieval system, without written permission from the publishers.

For permission to photocopy or use material electronically from this work, please access www.copyright.com (http://www.copyright.com/) or contact the Copyright Clearance Center, Inc. (CCC), 222 Rosewood Drive, Danvers, MA 01923, 978-750-8400. CCC is a not-for-profit organization that provides licenses and registration for a variety of users. For organizations that have been granted a photocopy license by the CCC, a separate system of payment has been arranged.

Trademark Notice: Product or corporate names may be trademarks or registered trademarks, and are used only for identification and explanation without intent to infringe.

Library of Congress Cataloging-in-Publication Data

Names: Chandrasekaran, Srinivasan, author.
Title: Advanced structural analysis with MATLAB / Srinivas Chandrasekaran.
Description: Boca Raton: Taylor & Francis, a CRC title, part of the Taylor & Francis imprint, a member of the Taylor & Francis Group, the academic division of T&F Informa, plc, 2018. | Includes bibliographical references.
Identifiers: LCCN 2018036253 | ISBN 9780367026455 (hardback : acid-free paper)
Subjects: LCSH: Structural analysis (Engineering)--Mathematics. | MATLAB.
Classification: LCC TA645 .C375 2018 | DDC 624.1/70285536--dc23
LC record available at https://lccn.loc.gov/2018036253

Visit the Taylor & Francis Web site at
http://www.taylorandfrancis.com

and the CRC Press Web site at
http://www.crcpress.com

Contents

Foreword vii
Preface ix
Author xi

Chapter 1 Planar Orthogonal Structures 1

    1.1   Introduction to Structural Analysis 1
    1.2   Indeterminacy 1
            1.2.1   Static Indeterminacy 2
            1.2.2   Kinematic Indeterminacy 2
                      1.2.2.1   Continuous Beam 3
                      1.2.2.2   Fixed Beam 3
                      1.2.2.3   Simply Supported Beam 4
                      1.2.2.4   Frame 4
    1.3   Linear Equations 5
            1.3.1   Inverse of a Matrix 6
            1.3.2   Solution of Linear Equations 7
    1.4   Matrix Operations 8
            1.4.1   Submatrix 8
            1.4.2   Partitioning of Matrix 10
            1.4.3   Cross-Partitioning of Matrix 10
            1.4.4   Banded Matrix 12
    1.5   Standard Beam Element 12
            1.5.1   Estimating Rotational Coefficients 19
    1.6   Beam Element with Varying Flexural Rigidity 22
    1.7   Planar Orthogonal Structures 27
    1.8   Example Problems 30
            1.8.1   Continuous Beam 30
            1.8.2   Computer Program for Continuous Beam 38
            1.8.3   Orthogonal Frame 42
            1.8.4   Computer Program for Orthogonal Frame 50
            1.8.5   Step Frame 55
            1.8.6   Computer Program for Step Frame 61

Chapter 2 Planar Non-Orthogonal Structures 67

    2.1   Planar Non-Orthogonal Structure 67
    2.2   Stiffness Matrix Formulation 69
    2.3   Transformation Matrix 71
    2.4   Transformation Matrix for End Moments 73
    2.5   Global Stiffness Matrix 76
    2.6   Important Steps in Analysis of Non-Orthogonal Structures 77

v

Chapter 3	Planar Truss Structures 123		
	3.1	Planar Truss System 123	
		3.1.1 Transformation Matrix 123	
		3.1.2 Stiffness Matrix 124	
Chapter 4	Three-Dimensional Analysis of Space Frames 159		
	4.1	Three-Dimensional Analysis of Structures 159	
	4.2	Beam Element 159	
	4.3	The Stiffness Matrix 160	
	4.4	Transformation Matrix 164	
	4.5	Member Rotation Matrix 166	
	4.6	Y-Z-X Transformation 168	
	4.7	Z-Y-X Transformation 172	
	4.8	The Ψ Angle 173	
	4.9	Analysis of Space Frame 174	
Chapter 5	Analysis of Special Members 205		
	5.1	Three-Dimensional Analysis of Truss Structures 205	
	5.2	Special Elements 207	
	5.3	Non-Prismatic Members 209	

Appendix 221

Bibliography 233

Index 237

Foreword

There has been tremendous growth in the field of civil engineering in recent times. The structures have evolved from being simple and straightforward in configuration to more complex ones in terms of shape, geometry and design requirements. Modern-day architecture demands very intricate structural design in making a built environment efficient and sustainable. In the era of fast-paced growth, there are numerous finite element packages available to reduce the computational effort and time. However, one should have a strong foothold in the basic concepts of structural engineering as most of the numerical work is handled by computer.

This book addresses the basic structural forms such as orthogonal and non-orthogonal planar frames, space frames and trusses that go into the making of complicated structures. The approach adopted is based on the intuition that an engineer's ability to perceive a concept is through simple models. Numerous examples have been included to illustrate the fundamental concepts more clearly. Hence, in addition to conceptual understanding, an effort is made to include the basics of computation with detailed examples. The book consists of five chapters.

The analysis steps have been explained in a classroom style of teaching and the computer programs for MATLAB® platform have been introduced in the form of examples. These computer programs cover matrix operation for a variety of structural forms and responses. The illustrative examples in the book enhance the understanding of the structural concepts stimulating interest in learning, creative thinking and design. In conclusion, the book stems from a void in conceptual understanding of the structural behavior, based on problem solving experience with students exposed to engineering mechanics and mechanics of materials.

The author of this book, Professor Srinivasan Chandrasekaran, is a renowned teacher and researcher with diverse industrial experience in structural engineering. He has already authored many peer-reviewed journal articles, conference papers, textbooks and reports on international projects. His rich expertise and experience in the teaching of fundamentals of structural analysis at IIT Madras has been brought out now in the form of this new book. I strongly believe this textbook to be an ideal resource for students and teachers, and a comprehensive reference for practitioners. I congratulate Professor Chandrasekaran for his total commitment to the advancement of technical education. I hope that many will learn from this book and apply its principles in their profession.

Katta Venkataramana
Professor, National Institute of Technology
Suratkal, India

Preface

Analysis of civil engineering structures is becoming more complex essentially due to different structural forms that are conceived by architects and engineers to accommodate various functional requirements. Conventional analysis tools guide engineering graduates and practicing professionals in addressing such issues, but accuracy and compactness, in terms of varied solutions, are difficult. Matrix methods in general, and the stiffness method in particular, are very powerful tools to model complicated structural forms and to perform the required analysis. However, the application of matrix algorithms in a more generic form to solve all types of problems, namely beams, trusses, planar orthogonal frames, planar non-orthogonal frames, three-dimensional trusses and space frames, needs to be addressed in a step-by-step manner to resolve all possible doubts that may arise during solution procedures. While acknowledging the ingenious efforts made by authors from all over the world on this front, this book is a humble attempt to revisit these concepts with more elaborate explanations and very strong hand-supportive computer coding. One of the main objectives of this book is to help solve problems using matrix methods along with a familiarization of computer coding to solve such problems. MATLAB® is a well-established and proven tool to handle such complex problems in a very simple and highly supportive manner.

This book starts with an analysis of beams and planar orthogonal frames, and it also addresses problems of truss elements, special elements, planar non-orthogonal frames, three-dimensional trusses and space frames. One of the most attractive features of this book is how it explains the problem solution which is highly compatible with computer coding using MATLAB. Each problem is carefully examined and degrees-of-freedom (both restrained and unrestrained) are marked in a more generic manner with a uniform sign convention throughout the text of this book. Matrix formulation of the problem is clearly presented step by step; this is also followed while writing the computer code for solving the problem. Example problems given in each chapter are solved using MATLAB coding, while input data to use the coding is explained in detail. The output obtained from the coding is plotted as a screenshot for better inference of results. Numerous exercise problems are given along with solutions that enable the readers to use the same computer code with a minor modification to suit the inputs for the problems.

One of the salient features of this book is that similar computer code as that for two-dimensional is used for three-dimensional analysis, except transformations that are required from local to global axes systems. The book also supports many practice papers to ensure a high level of confidence while solving such problems. A Solutions Manual and additional instructor resources are available as downloadable e-resources on the book's CRC Press webpage at https://www.crcpress.com/Advanced-Structural-Analysis-with-MATLAB/Chandrasekaran/p/book/9780367026455. MATLAB files are also available in downloadable format on same webpage. The author sincerely thanks the Centre for Continuing Education, Indian Institute of Technology (IIT) Madras, for extending administrative support in preparing the manuscript of this book. The author also thanks MATLAB for permitting usage of MATLAB codes throughout the text of this book.

The computer programs used in the book are written using MATLAB, following well-established programming concepts. Program coding is written following the same steps as those for conventional analysis using the stiffness method. The program codes were written by Nagavinothini.R and verified by a team of research scholars, Department of Ocean Engineering, IIT Madras. Nagavinothini.R is senior research scholar in the Department of Ocean Engineering at IIT Madras, Chennai, India. She is currently working on dynamic analysis of offshore new generation compliant platforms in ultra-deep waters under environmental and accidental loads. She is a University Rank Holder and Gold Medalist, who has published many research papers in refereed journals. Her research interests include dynamic analysis of structures, computer-aided analysis of structures, design and optimization of structures.

Utmost care has been taken to check solutions and to correct errors, but the author does not claim or guarantee the correctness of outputs using the provided computer codes. Readers are asked to verify based on their independent capacity and then use the codes for practical applications.

Srinivasan Chandrasekaran
Department of Ocean Engineering
Indian Institute of Technology
Madras, India

MATLAB® is a registered trademark of The MathWorks, Inc. For product information, please contact:

The MathWorks, Inc.
3 Apple Hill Drive
Natick, MA 01760-2098 USA
Tel: 508-647-7000
Fax: 508-647-7001
E-mail: info@mathworks.com
Web: www.mathworks.com

Author

Srinivasan Chandrasekaran is a professor in the Department of Ocean Engineering, Indian Institute of Technology Madras, India. He has more than 24 years of teaching, research and industrial experience, during which he has supervised many sponsored research projects and offshore consultancy assignments both in India and abroad. He has also been a visiting fellow at the University of Naples Federico II, Italy (MIUR Fellow), during which time he conducted research on advanced nonlinear modeling and analysis of structures under different environmental loads with experimental verifications. He has published approximately 130 research papers in international journals and refereed conferences organized by professional societies around the world. He has also authored textbooks, which are quite popular among graduate students in civil and ocean engineering. He is a member of many national and international professional bodies and delivered many invited lectures and keynote addresses at international conferences, workshops and seminars organized in India and abroad.

1 Planar Orthogonal Structures

1.1 INTRODUCTION TO STRUCTURAL ANALYSIS

The first and foremost step in structural analysis is problem formulation using an appropriate mathematical model. There are two models widely used in classical structural analysis, namely: (1) the statically determinate model and (2) the statically indeterminate model. Statically determinate models are relatively easier as they use only the basic equations of static equilibrium to solve the problem. Hence, one should look for ways to solve problems related to statically indeterminate models through computer methods. In order to solve the problems, it is important to formulate a standard procedure, which should be generic in nature and not problem-specific. Thus, it is important to note that the models should be restrained from any action to enable solution of the problem by using a standard equation of statics. This can be done by grouping the formation. Grouping is done through two methods, namely: (1) the flexibility method; and (2) the stiffness method.

Both of the previously mentioned methods are frequently used to simplify the model to be solved by using only the standard equation of statics. Both methods are equally powerful and useful; there is no supremacy of one method over the other. Any method can be used for grouping based on the user's convenience. But, a method that is easily programmable is preferred, as the main objective of this book is to make the problems solvable through computer methods. There is a significant difference in identifying the unknowns for formulating the problem. In the flexibility method, the unknowns are actions such as shear force, axial force and bending moment. In the stiffness method, however, the unknowns are displacements such as translational and rotational displacements. Basic assumptions applicable to both methods are as follows:

1. A linear relationship exists between an applied load and the resulting displacement of the structure. This makes the principle of superposition valid through the formulation.
2. The material of the structure must obey Hooke's Law, which says that the material must not be stressed beyond its elastic limit.
3. The equations of static equilibrium shall be developed using the geometry of the un-deflected model. The change in geometry caused by the imposed loads is negligible when compared to original geometry.

1.2 INDETERMINACY

Both flexibility and stiffness methods circumscribe the problem formulations around the term indeterminacy. It is important to understand indeterminacy in terms of problem formulation by either the flexibility or the stiffness method. There are two types of indeterminacy, namely:

1. Static indeterminacy
2. Kinematic indeterminacy

1.2.1 STATIC INDETERMINACY

Static indeterminacy is the term related to the flexibility approach. It is defined as the number of actions (e.g. shear force, axial force, bending moment) that can be either external or internal, that must be released in order to transform the structural system into a stable statically determinate system. Thus, the objective is to convert the known structural system into a statically determinate and stable system, for which the number of actions has to be identified. The *degree of static indeterminacy* is defined as the number of released actions, which specify the number of special independent equations that must be developed in terms of the released actions to analyze the system. So, the approach used in the formulation and solution is the *flexibility approach*.

1.2.2 KINEMATIC INDETERMINACY

Kinematic indeterminacy is the term related to the stiffness approach. It refers to the number of independent components of joint displacements (both translational and rotational) with respect to a specified coordinate axis that is required to describe the response of the system under any arbitrary load. It can be seen that the kinematic indeterminacy or stiffness method is trying to reach a generic solution. This problem formulation needs to identify the number of independent displacement components, which will be invoked under the external forces acting on the system of any nature. It is important to note that the structure must be restrained to convert or transform the system into a kinematically determinate structure. A structure with all joint displacements restrained is the formulation. The *degree of kinematic indeterminacy* is defined as the number of unrestrained components of the joint displacements (both rotational and translational). It is important to know that the degree of kinematic indeterminacy specifies the number of independent equations that must be written in terms of unrestrained displacements, if the system is to be analyzed using the stiffness approach.

The differences between flexibility and stiffness methods are summarized as follows:

Flexibility Approach	**Stiffness Approach**
This deals with static indeterminacy	This deals with kinematic indeterminacy
The unknowns are actions such as shear force, axial force, bending moments, etc.	The unknowns are joint displacements such as rotational and translational displacements
The problem formulation converts the structural system into a statically determinate structure	The problem formulation converts the structural system into a kinematically determinate structure

Planar Orthogonal Structures

Thus, the static and kinematic indeterminacies of any structural system are indicators of the amount or extent of computational effort required to analyze the structural system either using the flexibility approach or the stiffness approach. It is clear that the structural analysis can be carried out by any of the two methods, which are equally useful and powerful numerically. The unknowns are released to convert the structural system into a statically or kinematically determinate system, so that the standard equations can be used to solve the system under applied loads.

If the problem formulation reduces the number of unknowns, then it is the best formulation attempted by a mathematician or an engineer. The number of unknowns in the system of equations purely depends on the choice of the method demanded. For a computer method of structural analysis, one should keep in mind that the method recommended should be more or less generic and not problem-specific. The degree of static and kinematic indeterminacies of standard problems are given subsequently.

1.2.2.1 Continuous Beam

Consider a continuous beam of three spans with one hinged joint and three roller joints with reactions R1, R2, R3 and R4, as shown in Figure 1.1. The displacement unknowns, neglecting axial deformations are θ1, θ2, θ3 and θ4. We all know that there are three systems of standard equations available to solve the problem. The degree of static indeterminacy and the degree of kinematic indeterminacy are as follows:

$$\text{Degree of static indeterminacy} = \text{Number of unknown reactions} - \text{system of standard equations}$$
$$= 5 - 3 = 2.$$

$$\text{Degree of kinematic indeterminacy} = \text{Number of displacements}$$
$$= 4$$

1.2.2.2 Fixed Beam

Let us now consider a fixed beam with reactions R_1, R_2 and R_3 at support A. Similarly, the reactions at support B are R_4, R_5 and R_6, as shown in Figure 1.2. The rotational displacement which is free to move is zero. Thus,

Degree of static indeterminacy $= 6 - 3 = 3$
Degree of kinematic indeterminacy $= 0$

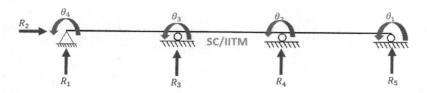

FIGURE 1.1 Continuous beam.

FIGURE 1.2 Fixed beam.

1.2.2.3 Simply Supported Beam

Let us consider a simply supported beam with reaction components R_1, R_2 and R_3. The displacements are θ_1 and θ_2 as shown in Figure 1.3. Thus,

Degree of static indeterminacy = 3 − 3 = 0
Degree of kinematic indeterminacy = 2

1.2.2.4 Frame

Let us consider a single story single bay frame with one end fixed and the other end on a roller support as shown in Figure 1.4. The unknown reactions are R_1, R_2, R_3,

FIGURE 1.3 Simply supported beam.

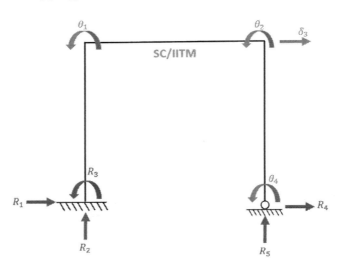

FIGURE 1.4 Single story single bay frame.

Planar Orthogonal Structures

R4 and R5. The independent displacements are θ_1, θ_2, δ_3 and δ_4 by neglecting axial deformation.

Degree of static indeterminacy $= 5 - 3 = 2$
Degree of kinematic indeterminacy $= 4$

From the previous examples, it can be seen that the number of unknowns is different depending on the choice of method. In order to summarize for the choice of method of analysis which governs the system of equations, the following is noteworthy:

1. There are essentially two methods to solve the statically indeterminate structures, such as the flexibility method and stiffness method. So, the choice of the method depends on the computational convenience.
2. For the flexibility method, there are several alternatives for the redundant or unknowns. Thus, the choice of the redundant has a significant effect on the computational effort.
3. On the other hand, there is no choice of unknown quantities in the stiffness method because there is only one possible restrained structure. Therefore, this has a set of standard procedures.
4. Based on computer methods of structural analysis, one can say that the choice of the method should not be geometry specific. It should be more generic and repetitive in nature.

Thus, the stiffness method is a better choice fulfilling all the previously mentioned requirements. In this book, the stiffness method is used for elaborating the application procedures. After identifying the variables in a given system, such as rotational or translational displacements for every joint, it will result in the formation of a set of linear equations.

1.3 LINEAR EQUATIONS

A system of 'm' linear equations with 'n' unknowns is expressed as follows:

$$a_{11}x_1 + a_{12}x_2 + \ldots + a_{1n}x_n = b_1$$

$$a_{21}x_1 + a_{22}x_2 + \ldots + a_{2n}x_n = b_2 \qquad (1.1)$$

$$\ldots \ldots$$

$$a_{m1}x_1 + a_{m2}x_2 + \ldots + a_{mn}x_n = b_m$$

The previous set of equations can be written in matrix form as follows:

$$\begin{bmatrix} a_{11} & a_{12} & a_{1n} \\ a_{21} & a_{22} & a_{2n} \\ a_{m1} & a_{m2} & a_{mn} \end{bmatrix} \begin{Bmatrix} x_1 \\ x_2 \\ x_n \end{Bmatrix} = \begin{Bmatrix} b_1 \\ b_2 \\ b_m \end{Bmatrix}$$

$$[A]\{x\} = \{B\} \qquad (1.2)$$

Pre-multiply the previous equation with A^{-1} on both sides,

$$[A]^{-1}[A]\{x\} = [A]^{-1}\{B\}$$

$$[I]\{x\} = [A]^{-1}\{B\}$$

$$\{x\} = [A]^{-1}\{B\} \qquad (1.3)$$

Thus, the previous equation gives the unknown 'x' by multiplying the inverse of the matrix A and B vector. Now, the problem is to compute the inverse of a matrix.

1.3.1 INVERSE OF A MATRIX

Inverse of matrix A is given by,

$$[A]^{-1} = \frac{adj\, A}{|A|} \qquad (1.4)$$

In a given square matrix, replace each element a_{ij} of the matrix $[A]$ by its cofactor α_{ij}. Transform the cofactor matrix to obtain adjoint matrix. The following simple example will give the procedure to find the inverse of matrix.

Consider a matrix, $A = \begin{bmatrix} 1 & 5 & 2 \\ 0 & 4 & 1 \\ 0 & 2 & 1 \end{bmatrix}$. Find $[A]^{-1}$ by adjoint method.

$$|A| = 1\{(4 \times 1) - (2 \times 1)\} = 2$$

Cofactors are given by,

$$\alpha_{11} = (-1)^{1+1} \begin{vmatrix} 4 & 1 \\ 2 & 1 \end{vmatrix} = 2$$

$$\alpha_{12} = (-1)^{1+2} \begin{vmatrix} 0 & 1 \\ 0 & 1 \end{vmatrix} = 0$$

$$\alpha_{13} = (-1)^{1+3} \begin{vmatrix} 0 & 4 \\ 0 & 2 \end{vmatrix} = 0$$

$$\alpha_{21} = (-1)^{1+3} \begin{vmatrix} 5 & 2 \\ 2 & 1 \end{vmatrix} = -1$$

Planar Orthogonal Structures

$$\alpha_{22} = (-1)^{2+2} \begin{vmatrix} 1 & 2 \\ 0 & 1 \end{vmatrix} = 1$$

$$\alpha_{23} = (-1)^{2+3} \begin{vmatrix} 1 & 5 \\ 0 & 2 \end{vmatrix} = -2$$

$$\alpha_{31} = (-1)^{3+1} \begin{vmatrix} 5 & 2 \\ 4 & 1 \end{vmatrix} = -3$$

$$\alpha_{32} = (-1)^{3+2} \begin{vmatrix} 1 & 2 \\ 0 & 1 \end{vmatrix} = -1$$

$$\alpha_{33} = (-1)^{3+3} \begin{vmatrix} 1 & 5 \\ 0 & 4 \end{vmatrix} = 4$$

The cofactor matrix is written as:

$$\alpha_{ij} = \begin{bmatrix} 2 & 0 & 0 \\ -1 & 1 & -2 \\ -3 & -1 & 4 \end{bmatrix}$$

$$\text{Adj } A = [\alpha_{ij}]^T = \begin{bmatrix} 2 & -4 & -3 \\ 0 & 1 & -1 \\ 0 & -2 & 4 \end{bmatrix}$$

Thus,

$$[A]^{-1} = \frac{\text{adj } A}{|A|} = \frac{1}{2}\begin{bmatrix} 2 & -4 & -3 \\ 0 & 1 & -1 \\ 0 & -2 & 4 \end{bmatrix} = \begin{bmatrix} 1 & 1/2 & -3/2 \\ 0 & 1/2 & -1/2 \\ 0 & -1 & 2 \end{bmatrix}$$

To check:

$$[A]^{-1}[A] = \begin{bmatrix} 1 & 1/2 & -3/2 \\ 0 & 1/2 & -1/2 \\ 0 & -1 & 2 \end{bmatrix} \times \begin{bmatrix} 1 & 5 & 2 \\ 0 & 4 & 1 \\ 0 & 2 & 1 \end{bmatrix} = [I] = \begin{bmatrix} 1 & 0 & 0 \\ 0 & 1 & 0 \\ 0 & 0 & 1 \end{bmatrix}$$

1.3.2 Solution of Linear Equations

Let us express the matrix A as a system of equations,

$$x_1 + 5x_2 + 2x_3 = 2$$

$$4x_2 + x_3 = 5$$

$$2x_2 + x_3 = 4$$

Now, these set of equations have to be solved to get the variables (x_1, x_2, x_3). In matrix form, the equations can be written as follows:

$$\begin{bmatrix} 1 & 5 & 2 \\ 0 & 4 & 1 \\ 0 & 2 & 1 \end{bmatrix} \begin{Bmatrix} x_1 \\ x_2 \\ x_3 \end{Bmatrix} = \begin{Bmatrix} 2 \\ 5 \\ 4 \end{Bmatrix}$$

$$\begin{Bmatrix} x_1 \\ x_2 \\ x_3 \end{Bmatrix} = [A]^{-1} \begin{Bmatrix} 2 \\ 5 \\ 4 \end{Bmatrix} = \begin{bmatrix} 1 & 1/2 & -3/2 \\ 0 & 1/2 & -1/2 \\ 0 & -1 & 2 \end{bmatrix} \begin{Bmatrix} 2 \\ 5 \\ 4 \end{Bmatrix} = \begin{Bmatrix} -6.5 \\ 0.5 \\ 3 \end{Bmatrix}$$

If one can generate a system of equations with unknowns as variables, then this set of equations can be solved using matrix inversion. This is an easy method to solve for the variable as given by equation 1.3.

$$\{x\} = [A]^{-1} \{B\}$$

This is true only when $[A]^{-1}$ exists. It should also be noted that $\{x\}$ purely depends on $\{B\}$ and $[A]^{-1}$ does not change to get the value of $\{x\}$. Assume matrix A as a stiffness matrix of a given system, B as a load vector and x as a displacement vector. Through this comparison, it can be seen that the value of the displacement vector for a changed load vector can be found without changing the inverse of the stiffness matrix. In the case that $\{B\}$ is zero and when $[A]^{-1}$ also exists, then the possible solution is said to be a trivial solution, that is, $x=0$. In this case, $[A]^{-1}$ does not exist, then the previous set of equations will lead to non-trivial solution.

1.4 MATRIX OPERATIONS

1.4.1 SUBMATRIX

Let A be the given matrix, then the submatrix is defined as a matrix formed by deleting specified rows and columns of the matrix A. Instead of deleting the rows and columns, partitioning can also be done. This is a useful technique when the matrix size is very large.

Let us assume a set of algebraic equations as follows:

$$\begin{aligned} y_1 &= a_{11}x_1 + a_{12}x_2 + \ldots + a_{1q}x_q + a_{1,q+1}x_{q+1} + \ldots + a_{1n}x_n \\ y_2 &= a_{21}x_1 + a_{22}x_2 + \ldots + a_{2q}x_q + a_{2,q+1}x_{q+1} + \ldots + a_{2n}x_n \\ &\quad \ldots \\ &\quad \ldots \\ &\quad \ldots \\ y_n &= a_{n1}x_1 + a_{n2}x_2 + \ldots + a_{nq}x_q + a_{n,q+1}x_{q+1} + \ldots + a_{nn}x_n \end{aligned} \quad (1.5)$$

Planar Orthogonal Structures

Then, the previous set of equations can also be grouped.

$$\begin{aligned}
y_1 &= \left(a_{11}x_1 + a_{12}x_2 + \ldots + a_{1q}x_q\right) + \left(a_{1,q+1}x_{q+1} + \ldots + a_{1n}x_n\right) \\
y_2 &= \left(a_{21}x_1 + a_{22}x_2 + \ldots + a_{2q}x_q\right) + \left(a_{2,q+1}x_{q+1} + \ldots + a_{2n}x_n\right) \\
&\quad \ldots \\
y_n &= \left(a_{n1}x_1 + a_{n2}x_2 + \ldots + a_{nq}x_q\right) + \left(a_{n,q+1}x_{q+1} + \ldots + a_{nn}x_n\right)
\end{aligned} \quad (1.6)$$

Now let us express both sets of equations in matrix form:

$$\begin{Bmatrix} y_1 \\ y_2 \\ y_n \end{Bmatrix} = \begin{bmatrix} a_{11} & a_{12} & a_{1q} \\ a_{21} & a_{22} & a_{2q} \\ a_{n1} & a_{n2} & a_{nq} \end{bmatrix} \begin{Bmatrix} x_1 \\ x_2 \\ x_q \end{Bmatrix} + \begin{bmatrix} a_{1,q+1} & \ldots & a_{1n} \\ a_{2,q+2} & \ldots & a_{2n} \\ a_{n,q+1} & \ldots & a_{nn} \end{bmatrix} \begin{Bmatrix} x_{q+1} \\ x_{q+2} \\ x_n \end{Bmatrix} \quad (1.7)$$

Now the vector y can be written as,

$$\{y\} = [A_1]\{x_1\} + [A_2]\{x_2\} \quad (1.8)$$

where,

$$[A_1] = \begin{bmatrix} a_{11} & a_{12} & a_{1q} \\ a_{21} & a_{22} & a_{2q} \\ a_{n1} & a_{n2} & a_{nq} \end{bmatrix}$$

$$[A_2] = \begin{bmatrix} a_{1,q+1} & \ldots & a_{1n} \\ a_{2,q+2} & \ldots & a_{2n} \\ a_{n,q+1} & \ldots & a_{nn} \end{bmatrix}$$

$$\{x_1\} = \begin{Bmatrix} x_1 \\ x_2 \\ x_q \end{Bmatrix}$$

$$\{x_2\} = \begin{Bmatrix} x_{q+1} \\ x_{q+2} \\ x_n \end{Bmatrix}$$

There should be a perfect compatibility among the multiplying matrices, shown as follows:

$$\{y\}_{n\times 1} = [A_1]_{n\times q}\{x_1\}_{q\times 1} + [A_2]_{n\times (n-q)}\{x_2\}_{(n-q)\times 1} \quad (1.9)$$

The number of columns and the number of rows of the adjacent multipliers should be same. The compatibility is required to ensure grouping. Now, it can be said that $[A_1]$ is a submatrix of $[A]$ of size $n \times q$ and $[A_2]$ is a submatrix of $[A]$ of size $n \times (n-q)$.

1.4.2 Partitioning of Matrix

Let,

$$\{y\} = [A]\{x\} \quad (1.10)$$

After partitioning,

$$\{y\} = \begin{bmatrix} [A_1] & | & [A_2] \end{bmatrix} \begin{Bmatrix} \{x_1\} \\ - \\ \{x_2\} \end{Bmatrix} \quad (1.11)$$

Thus,

$$\{y\} = [A_1]\{x_1\} + [A_2]\{x_2\} \quad (1.12)$$

The previous equation is called partitioned matrix. Matrix $[A]$ is vertically partitioned and vector $\{x\}$ is horizontally partitioned. To make the valid partition of $[A]$ and $\{x\}$, it is important to establish compatibility; that is, the number of columns of $[A_1]$ must correspond to the number of rows of $\{x_1\}$, to make $[A_1]\{x_1\}$ valid.

1.4.3 Cross-Partitioning of Matrix

$$\{y\} = [A]\{x\}$$

Let $[A]$ be partitioned both horizontally and vertically into submatrices,

$$[A] = \begin{bmatrix} [A_{11}]_{p \times q} & [A_{12}]_{p \times (n-q)} \\ [A_{21}]_{(m-p) \times q} & [A_{22}]_{(m-p) \times (n-q)} \end{bmatrix}_{m \times n} \quad (1.13)$$

Let $\{x\}$ also be portioned horizontally,

$$\{x\}_{n \times 1} = \begin{Bmatrix} \{x_1\}_{q \times 1} \\ \{x_2\}_{(n-q) \times 1} \end{Bmatrix}_{n \times 1} \quad (1.14)$$

Therefore, the resulting matrix $\{y\}$ will also be a horizontally partitioned matrix.

$$\{y\}_{m \times 1} = \begin{Bmatrix} \{y_1\}_{p \times 1} \\ \{y_2\}_{(m-p) \times 1} \end{Bmatrix}_{m \times 1} \quad (1.15)$$

Planar Orthogonal Structures

Therefore,

$$\left\{\begin{array}{c}\{y_1\} \\ - \\ \{y_2\}\end{array}\right\} = \left[\begin{array}{c|c}[A_{11}] & [A_{12}] \\ \hline [A_{21}] & [A_{22}]\end{array}\right] \left\{\begin{array}{c}\{x_1\} \\ - \\ \{x_2\}\end{array}\right\} \quad (1.16)$$

This means that the matrix [A], which has both a horizontal and vertical partitioning, is called a cross-portioned matrix.

Once portioning is done, the following equations are valid.

$$\{y_1\} = [A_{11}]\{x_1\} + [A_{12}]\{x_2\}$$

$$\{y_2\} = [A_{21}]\{x_1\} + [A_{22}]\{x_2\} \quad (1.17)$$

The inverse is also valid for a partitioned matrix, which is very advantageous.

Let [A] be the following matrix with horizontal and vertical partitioning,

$$A = \begin{bmatrix} 1 & 4 & 0 & 0 \\ 2 & 2 & 0 & 0 \\ 0 & 0 & 3 & -1 \\ 0 & 0 & -5 & 2 \end{bmatrix}$$

Now, [A] can be written as,

$$A = \begin{bmatrix} [A_{11}] & [A_{12}] \\ [A_{21}] & [A_{22}] \end{bmatrix}$$

It can also be written as,

$$A = \begin{bmatrix} [A_{11}] & 0 \\ 0 & [A_{22}] \end{bmatrix}$$

Now, A^{-1} can also be expressed as a set of submatrices as follows:

$$[A]^{-1} = \begin{bmatrix} [B_{11}] & [B_{12}] \\ [B_{21}] & [B_{22}] \end{bmatrix}$$

where,

$[B_{12}]$ = Inverse of $[A_{12}] = 0$

$[B_{21}]$ = Inverse of $[A_{21}] = 0$

$[B_{11}]$ = Inverse of $[A_{11}] = \dfrac{1}{-6}\begin{bmatrix} 2 & -4 \\ -2 & 1 \end{bmatrix}$

$$[B_{22}] = \text{Inverse of } [A_{22}] = \begin{bmatrix} 2 & 1 \\ 5 & 3 \end{bmatrix}$$

Now the advantage is that $[A]^{-1}$ can be easily written as:

$$[A]^{-1} = \begin{bmatrix} -1/3 & 2/3 & 0 & 0 \\ 1/3 & -1/6 & 0 & 0 \\ 0 & 0 & 2 & 1 \\ 0 & 0 & 5 & 3 \end{bmatrix}$$

Thus, determining the inverse of the 4×4 matrix is made easier with cross-partitioning of the matrix. Partitioning benefits inverting a 2×2 matrix, instead of a 4×4. This can lead to substantial savings in time and computational efforts.

1.4.4 Banded Matrix

Matrices in structural analysis show certain special properties. The matrices are real, symmetric, positive definite and banded. They can be utilized for solving a large system of equations. Given matrix [A] is said to be positive definite only when the following condition is satisfied.

$$X^T A X > 0, \quad \text{for all non-zero column matrix of } \{x\}$$

For example, consider the following matrix:

$$A = \begin{bmatrix} a_{11} & a_{12} & 0 & 0 & 0 & 0 \\ a_{21} & a_{22} & a_{23} & 0 & 0 & 0 \\ 0 & a_{32} & a_{33} & a_{34} & 0 & 0 \\ 0 & 0 & a_{43} & a_{44} & a_{45} & 0 \\ 0 & 0 & 0 & a_{54} & a_{55} & a_{56} \\ 0 & 0 & 0 & 0 & a_{65} & a_{66} \end{bmatrix}$$

[A] is said to be a banded matrix with width $(2m+1)$ if all elements of a_{ij} for which $|i-j| > m$ are zero. For $m = 1$, the band width of the previously mentioned matrix is 3.

1.5 STANDARD BEAM ELEMENT

A beam element is one of the basic elements to be used in the structural analysis of planar orthogonal frames. There are some sign conventions that need to be followed before deriving the stiffness matrix of the beam element. The anticlockwise end moment, joint rotation and joint moments are taken as positive; the upward force or displacement of the joint is positive; the force toward the right or axial displacement toward the right of the joint is positive; the upward end shear at the ends of the beam is positive; the right direction force at the ends of the beam is positive. Let us limit

Planar Orthogonal Structures

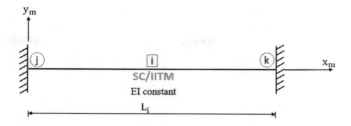

FIGURE 1.5 Standard beam element.

the discussion to planar orthogonal structures, which have either horizontal or vertical members.

A fixed beam is considered as a basic model. Consider a fixed beam undergoing deformation due to bending, neglecting the axial deformation. The standard fixed beam is shown in Figure 1.5. The two joints of the beam element are 'j' and 'k', and the length of the member is L_i. In terms of mathematical conditions, the considered beam element is fixed at nodes 'j' and 'k'; it has constant EI over its entire length. The left end of the beam is designated as the jth node and the right end of the beam is designated as the kth node. The member is designated as the ith member. (x_m, y_m) are local axes of the member. It is very important to note the axis system. The axis system is such that it has an origin at the jth end; x_m is directed toward the kth end. y_m is counterclockwise 90 degrees to the x_m axis. Therefore, the (x_m, y_m) plane defines the plane of bending the beam element. This is the conventional way of explaining the standard beam element. Let us neglect the axial deformation. For the stiffness method, one should identify possible displacements, both translational and rotational, at each end of the beam. So, the possible rotational and translational moments are shown in Figure 1.6. Suitable subscripts are used to denote the rotational and translational moments. Note the order by which the moments are marked, which allow the readers to understand the computer programs easily. The displacements at the jth end and the kth end are $\left(\theta_p, \delta_r\right)$ and $\left(\theta_q, \delta_s\right)$ respectively. All these displacements happen in the x_m, y_m plane and there is no out-of-plane bending.

Now, let us derive the stiffness coefficient, k_{ij}. According to classical definition, k_{ij} is the force in the ith degree-of-freedom by imposing unit displacement, which can

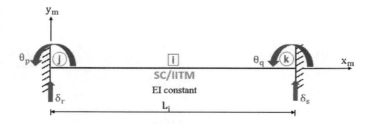

FIGURE 1.6 Rotational and translational moments in a beam element.

be either translational or rotational in the *j*th degree-of-freedom by keeping all other degrees-of-freedom restrained. There are two degrees-of-freedom:

1. Static degree-of-freedom
2. Kinematic degree-of-freedom

Static degree-of-freedom is related to the release of actions like shear force, bending moment, axial force, and so on. It is associated with the flexibility approach. Kinematic degree-of-freedom is related to displacements. It is associated with the stiffness approach. The degrees-of-freedom mentioned in the definition of stiffness are related to displacement and hence kinematic degree-of-freedom. In the previously mentioned, there are four degrees -of-freedom (two rotations and two translations). One should give unit displacement in each degree-of-freedom to find the forces at the respective degrees by keeping the remaining degrees-of-freedom restrained. k_{ij} is also defined as the moment in the *i*th degree-of-freedom by imposing unit rotation at the *j*th degree-of-freedom by keeping all other degrees-of-freedom restrained. Imposing unit displacement represents $\delta_r = 1$ or $\delta_s = 1$ and unit rotation implies $\theta_p = 1$ or $\theta_q = 1$.

Let us give unit rotation at the *j*th end, keeping all other degrees-of-freedom restrained, as shown in Figure 1.7. This will invoke members with end forces; $k_{pp}^i, k_{qp}^i, k_{rp}^i, k_{sp}^i$. k_{pp}^i is the force in the *p*th degree-of-freedom by giving unit displacement in the *p*th degree-of-freedom in the *i*th member. Similarly, k_{qp}^i is the force in the *q*th degree-of-freedom by giving unit displacement in the *p*th degree-of-freedom in the *i*th member.

The second subscript in all the notations is common; it is '*p*'. It indicates that the unit displacement is given at the *p*th degree. The stiffness coefficients are generated column-wise. Thus, the obtained stiffness coefficients correspond to the first column of the stiffness matrix. Similarly, apply unit rotation at the end *k* of the member, as shown in Figure 1.8. The stiffness coefficients in this case will be $k_{pq}^i, k_{qq}^i, k_{rq}^i, k_{sq}^i$. The stiffness coefficients are again obtained by applying unit displacements at the *j*th end and *k*th end, as shown in Figures 1.9 and 1.10. By applying unit displacement at the *j*th end, the stiffness coefficients are $k_{pr}^i, k_{qr}^i, k_{rr}^i, k_{sr}^i$. By applying unit displacement at the *k*th end, the following stiffness coefficients are obtained: $k_{ps}^i, k_{qs}^i, k_{rs}^i, k_{ss}^i$. A tangent can be drawn by connecting the deflected position of the beam at which the unit rotation is applied and the initial position of the beam at the other end. From

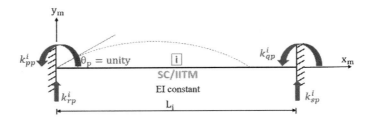

FIGURE 1.7 Fixed beam element – unit rotation at *j*th end.

Planar Orthogonal Structures

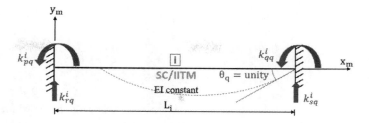

FIGURE 1.8 Fixed beam element – unit rotation at *k*th end.

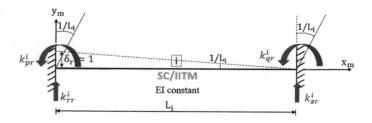

FIGURE 1.9 Fixed beam element – unit displacement at *j*th end.

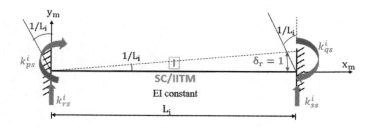

FIGURE 1.10 Fixed beam element – unit displacement at *k*th end.

which, one can say that the beam has undergone a rotation of $1/L_i$, where L_i is the length of the member. The rotation at the ends of the beam is equal to $1/L_i$.

For the *i*th beam element experiencing arbitrary end displacements, namely (θ_p, δ_r) and (θ_q, δ_s), corresponding end reactions (moment, shear) are required to be estimated. They need to be estimated by maintaining the equilibrium of the restrained member. The governing equations are as follows:

$$m_p^i = k_{pp}^i \theta_p + k_{pq}^i \theta_q + k_{pr}^i \delta_r + k_{ps}^i \delta_s \tag{1.18}$$

It can be seen from the previous equation that the first subscript of the equation corresponds to the end at which the unit rotation or displacement is applied. Similarly,

$$m_q^i = k_{qp}^i \theta_p + k_{qq}^i \theta_q + k_{qr}^i \delta_r + k_{qs}^i \delta_s \tag{1.19}$$

The force equations are,

$$p_r^i = k_{rp}^i \theta_p + k_{rq}^i \theta_q + k_{rr}^i \delta_r + k_{rs}^i \delta_s$$

$$p_s^i = k_{sp}^i \theta_p + k_{sq}^i \theta_q + k_{sr}^i \delta_r + k_{ss}^i \delta_s \quad (1.20)$$

The previous equations give the end moments and end shear forces for arbitrary displacements $\theta_p, \theta_q, \delta_r$ and δ_s, which are unity at respective degrees-of-freedom. These equations can be generalized by the following equation:

$$\{m_i\} = [k]_i \{\delta_i\} \quad (1.21)$$

where,

$$\{m_i\} = \begin{Bmatrix} m_p \\ m_q \\ p_r \\ p_s \end{Bmatrix}, \{\delta_i\} = \begin{Bmatrix} \theta_p \\ \theta_q \\ \delta_r \\ \delta_s \end{Bmatrix} \text{ and}$$

$$[k]_i = \begin{bmatrix} k_{pp} & k_{pq} & k_{pr} & k_{ps} \\ k_{qp} & k_{qq} & k_{qr} & k_{qs} \\ k_{rp} & k_{rq} & k_{rr} & k_{rs} \\ k_{sp} & k_{sq} & k_{sr} & k_{ss} \end{bmatrix}$$

We need to evaluate only a set of rotational coefficients in the stiffness matrix, and the other coefficients can be easily written in terms of the same rotational coefficients. These rotational coefficients are $k_{pp}^i, k_{pq}^i, k_{qp}^i, k_{qq}^i$. For example, in order to evaluate the end shear,

$$k_{rp}^i = \frac{k_{pp}^i + k_{qp}^i}{L_i}$$

$$k_{sp}^i = -\frac{k_{pp}^i + k_{qp}^i}{L_i} \quad (1.22)$$

The negative sign in the previous equation is due to the fact that the direction of k_{sp}^i is opposite to the end shear developed by the restraining moments as shown in Figure 1.11. In the fixed beam element with unit rotation at the end p, the moments developed at the ends p and q to control the applied unit rotation are k_{pp}^i and k_{qp}^i respectively. The anticlockwise moment developed on the beam will be equal to $k_{pp}^i + k_{qp}^i$. This anticlockwise moment has to be counteracted by the shear, which will be creating a clockwise couple in the beam element. The shear forces at the ends p and q are found to be the same in magnitude, which is given by $\dfrac{k_{pp}^i + k_{qp}^i}{L_i}$.

Planar Orthogonal Structures

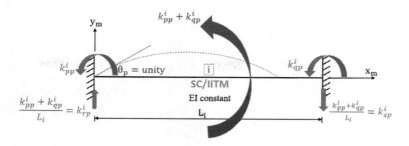

FIGURE 1.11 Fixed beam element – rotation coefficients.

The upward shear is positive, and the downward shear is negative, from which the coefficients k_{rp} and k_{sp} can be obtained. It can again be seen that the second subscript in the stiffness coefficients indicates the end at which the unit rotation is applied, and the first subscript indicates the respective forces in the degrees-of-freedom. In the same manner, the remaining stiffness coefficients can also be expressed in term of the rotation coefficients k_{pp}^i, k_{pq}^i, k_{qp}^i, k_{qq}^i. Thus, out of sixteen coefficients in the stiffness matrix, we need to evaluate only four coefficients.

Similarly, referring to Figure 1.8, where unit rotation is applied at the kth end, the rotational coefficients from equation 1.22 are given subsequently:

$$k_{rq}^i = \frac{k_{pq}^i + k_{qq}^i}{L_i}$$

$$k_{sq}^i = -\frac{k_{pq}^i + k_{qq}^i}{L_i}$$

By referring to Figure 1.9, where unit displacement is applied at the end j,

$$k_{pr}^i = \frac{k_{pp}^i + k_{pq}^i}{L_i}$$

$$k_{qr}^i = -\frac{k_{qp}^i + k_{qq}^i}{L_i} \qquad (1.23)$$

$$k_{rr}^i = \frac{k_{pr}^i + k_{qr}^i}{L_i} = \left[\frac{k_{pp}^i + k_{pq}^i}{(L_i)^2}\right] + \left[\frac{k_{qp}^i + k_{qq}^i}{(L_i)^2}\right] \qquad (1.24)$$

Thus,

$$k_{rr}^i = \frac{k_{pp}^i + k_{pq}^i + k_{qp}^i + k_{qq}^i}{(L_i)^2} \qquad (1.25)$$

Similarly,

$$k_{sr}^i = -\frac{k_{pp}^i + k_{pq}^i + k_{qp}^i + k_{qq}^i}{(L_i)^2} \tag{1.26}$$

By referring to Figure 1.10, where unit displacement is applied at the end k,

$$k_{ps}^i = -\frac{k_{pp}^i + k_{pq}^i}{L_i}$$

$$k_{qs}^i = -\frac{k_{qp}^i + k_{qq}^i}{L_i} \tag{1.27}$$

$$k_{rs}^i = -\frac{k_{ps}^i + k_{qs}^i}{L_i} = -\frac{k_{pp}^i + k_{pq}^i + k_{qp}^i + k_{qq}^i}{(L_i)^2} \tag{1.28}$$

$$k_{ss}^i = -\frac{k_{pp}^i + k_{pq}^i + k_{qp}^i + k_{qq}^i}{(L_i)^2} \tag{1.29}$$

Form the previous equations, it can be seen that the end shear coefficients are expressed in terms of rotational coefficients. Thus, by evaluating the four rotational coefficients, the stiffness matrix with sixteen stiffness coefficients can be framed. Thus, the stiffness matrix is given by

$$[k] = \begin{bmatrix} k_{pp} & k_{pq} & \frac{k_{pp}+k_{pq}}{L} & -\left(\frac{k_{pp}+k_{pq}}{L}\right) \\ k_{qp} & k_{qq} & \frac{k_{qp}+k_{qq}}{L} & -\left(\frac{k_{qp}+k_{qq}}{L}\right) \\ \frac{k_{pp}+k_{pq}}{L} & \frac{k_{pq}+k_{qq}}{L} & \frac{k_{pp}+k_{pq}+k_{qp}+k_{qq}}{L^2} & \frac{k_{pp}+k_{pq}+k_{qp}+k_{qq}}{L^2} \\ -\left(\frac{k_{pp}+k_{pq}}{L}\right) & -\left(\frac{k_{pq}+k_{qq}}{L}\right) & -\left(\frac{k_{pp}+k_{pq}+k_{qp}+k_{qq}}{L^2}\right) & -\left(\frac{k_{pp}+k_{pq}+k_{qp}+k_{qq}}{L^2}\right) \end{bmatrix}$$

Flexibility and stiffness matrices are related as follows:

$$[D][k] = [I] \tag{1.30}$$

Let $\begin{bmatrix} \delta_{jj} & \delta_{jk} \\ \delta_{kj} & \delta_{kk} \end{bmatrix}$ be the flexibility coefficients of a member with nodes j and k and $\begin{bmatrix} k_{pp} & k_{pq} \\ k_{qp} & k_{qq} \end{bmatrix}$ be the stiffness coefficients, then

Planar Orthogonal Structures

$$\begin{bmatrix} \delta_{jj} & \delta_{jk} \\ \delta_{kj} & \delta_{kk} \end{bmatrix} \begin{bmatrix} k_{pp} & k_{pq} \\ k_{qp} & k_{qq} \end{bmatrix} = \begin{bmatrix} 1 & 0 \\ 0 & 1 \end{bmatrix}$$

Our aim is to calculate the stiffness matrix [k]. This can be evaluated in the following way:

$$[k] = [D]^{-1}[I] \quad (1.31)$$

Now, let us evaluate the flexibility coefficients to form the flexibility matrix. Then, the stiffness matrix is derived from the flexibility matrix by the previous equation.

1.5.1 Estimating Rotational Coefficients

Consider a simply supported beam as shown in Figure 1.12. Unit rotation is applied at the *j*th end and the flexibility coefficients are determined. Similarly, flexibility coefficients are determined by applying unit displacement at the *k*th end, as shown in Figure 1.13. The flexibility coefficients $\left(\delta_{jj}^i, \delta_{kj}^i\right)$ define rotations at end *j* and *k* respectively of the *i*th member, caused due to unit moment applied at the *j*th end. Similarly, flexibility coefficients $\left(\delta_{jk}^i, \delta_{kk}^i\right)$ define rotations at *j*th and *k*th ends of the *i*th member due to unit moment applied at the *k*th end.

Let us consider a beam fixed at the end *q* and imposed by unit rotation at the end *p*, as shown in Figure 1.14. Similarly, assume the beam fixed at end *p* and imposed by unit rotation at end *q*, as shown in Figure 1.15. The stiffness coefficients $\left(k_{pp}^i, k_{qp}^i\right)$ define end moments required at *j*th and *k*th ends to maintain equilibrium when the *j*th end is subjected to unit rotation, while the *k*th end is restrained. Similarly, the

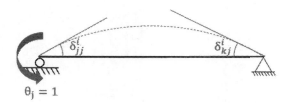

FIGURE 1.12 Simply supported beam – unit rotation at *j*.

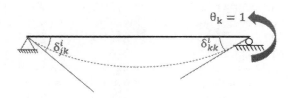

FIGURE 1.13 Simply supported beam – unit rotation at *k*.

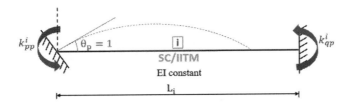

FIGURE 1.14 Fixed beam – unit rotation at *p*.

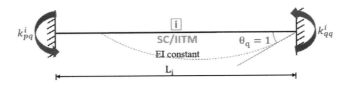

FIGURE 1.15 Fixed beam – unit rotation at *q*.

stiffness coefficients $\left(k_{pq}^i, k_{qq}^i\right)$ define end moments required at *j*th and *k*th ends to maintain equilibrium, when the *k*th end is subjected to unit rotation and the *j*th end is restrained.

Thus,

$$\begin{bmatrix} \delta_{jj} & \delta_{jk} \\ \delta_{kj} & \delta_{kk} \end{bmatrix} \begin{bmatrix} k_{pp} & k_{pq} \\ k_{qp} & k_{qq} \end{bmatrix} = \begin{bmatrix} 1 & 0 \\ 0 & 1 \end{bmatrix} \quad (1.32)$$

Expanding the previous equation,

$$k_{pp}^i \delta_{jj}^i + k_{qp}^i \delta_{jk}^i = 1$$

$$k_{pp}^i \delta_{kj}^i + k_{qp}^i \delta_{kk}^i = 1$$

$$k_{pq}^i \delta_{jj}^i + k_{qq}^i \delta_{jk}^i = 1$$

$$k_{pq}^i \delta_{kj}^i + k_{qq}^i \delta_{kk}^i = 1 \quad (1.33)$$

Let us denote the flexibility matrix as $[D_r]$ and stiffness matrix as $[k_r]$. The subscript *r* stands for rotational degrees-of-freedom. In order to estimate the flexibility matrix for the beam element, assume the simply supported beam with unit moment at the *j*th end, as shown in Figure 1.16. The anticlockwise moment is balanced by the clockwise couple created by the forces. The bending moment diagram is also shown in Figure 1.16 with tension at the top and compression at the bottom.

Let us replace the loading diagram with a conjugate beam, as shown in Figure 1.17. Taking the moment about *A*,

Planar Orthogonal Structures

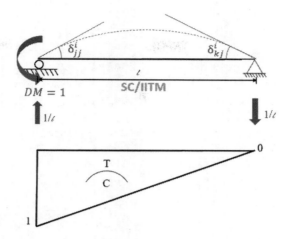

FIGURE 1.16 Simply supported beam.

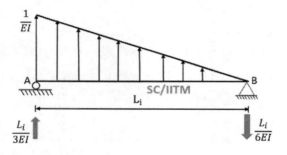

FIGURE 1.17 Conjugate beam.

$$V_B = \left[\left\{\frac{1}{2}L_i\left(\frac{1}{EI}\right)\right\}\frac{1}{3}L_i\right]\frac{1}{L_i} = \frac{L_i}{6EI} \text{ (downward)}$$

$$V_A = \left\{\frac{1}{2}L_i\left(\frac{1}{EI}\right)\right\} - \frac{L_i}{6EI} = \frac{L_i}{3EI} \text{ (upward)}$$

The same procedure is followed for the other case to derive the following flexibility matrix:

$$D_r = \begin{bmatrix} \dfrac{L}{3EI} & -\dfrac{L}{6EI} \\ -\dfrac{L}{6EI} & \dfrac{L}{3EI} \end{bmatrix}$$

Therefore,

$$k_r = [D_r]^{-1} = \frac{12(EI)^2}{L^2}\begin{bmatrix} \frac{L}{3EI} & \frac{L}{6EI} \\ \frac{L}{6EI} & \frac{L}{3EI} \end{bmatrix} = \begin{bmatrix} \frac{4EI}{L} & \frac{2EI}{L} \\ \frac{2EI}{L} & \frac{4EI}{L} \end{bmatrix}$$

Thus, from the previously mentioned four rotational coefficients, the whole stiffness matrix can be derived.

$$K_i = \begin{bmatrix} \frac{4EI}{l} & \frac{2EI}{l} & \frac{6EI}{l^2} & -\frac{6EI}{l^2} \\ \frac{2EI}{l} & \frac{4EI}{l} & \frac{6EI}{l^2} & -\frac{6EI}{l^2} \\ \frac{6EI}{l^2} & \frac{6EI}{l^2} & \frac{12EI}{l^3} & -\frac{12EI}{l^3} \\ -\frac{6EI}{l^2} & -\frac{6EI}{l^2} & -\frac{12EI}{l^3} & \frac{12EI}{l^3} \end{bmatrix}$$

1.6 BEAM ELEMENT WITH VARYING FLEXURAL RIGIDITY

Consider a beam with varying depth, as shown in Figure 1.18. The length of the beam is 5 m. Since the depth of the beam is varying, the flexural rigidity of the beam will also vary, even under constant material (E is constant). The moment of inertia will vary.

The stiffness matrix is developed by neglecting the axial deformation. Let us divide the beam into ten parts, as shown in Figure 1.19. A different moment of inertia for each part is assigned. For the cross-section of the beam at the end with 400 mm depth, the moment of inertia is given by,

$$I = \frac{300 \times 400^3}{12} = 1.6 \times 10^9 \text{ mm}^4$$

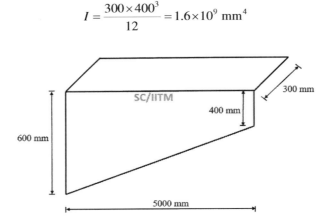

FIGURE 1.18 Beam with varying depth.

Planar Orthogonal Structures

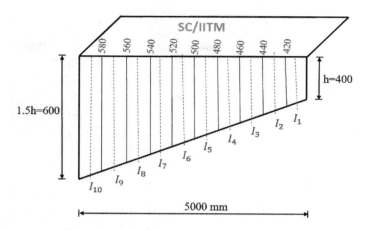

FIGURE 1.19 Beam with varying moment of inertia.

Now, the moment of inertia of each strip can be obtained by calculating the depth of the beam at each and every strip. We also know the average thickness of all the strips. Let h=400, then 1.5h=600. In the same way, the average thickness of all the strips can be expressed in terms of 'h'. The values are given in Table 1.1. Then, the moment of inertia of all the strips can be easily found. For example, the moment of inertia of strip 1 is given by,

$$I_1 = (1.025)^3 I = 1.077I$$

Similarly, the moment of inertia of all the strips can be calculated and the values are listed in Table 1.1.

The next step is to compute the loading diagram. Consider a simply supported beam to which a unit moment is applied in order to derive the flexibility coefficients. The deflected profile of the beam with flexibility coefficients is shown in Figure 1.20.

TABLE 1.1
M/EI Ordinates of Strips of the Beam-Unit Rotation at j

Strip	Average Thickness of Strips	Moment of Inertia (I)	Average Moment (M)	M/EI Ordinates
I_1	410 = 1.025h	1.077I	0.05	0.046/EI
I_2	430 = 1.075h	1.242I	0.15	0.121/EI
I_3	450 = 1.125h	1.424I	0.25	0.176/EI
I_4	470 = 1.175h	1.622I	0.35	0.216/EI
I_5	490 = 1.225h	1.838I	0.45	0.245/EI
I_6	510 = 1.275h	2.073I	0.55	0.265/EI
I_7	530 = 1.325h	2.326I	0.65	0.279/EI
I_8	550 = 1.375h	2.60I	0.75	0.288/EI
I_9	570 = 1.425h	2.894I	0.85	0.294/EI
I_{10}	590 = 1.475h	3.209I	0.95	0.296/EI

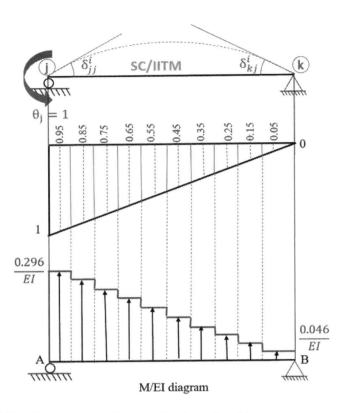

FIGURE 1.20 Simply supported beam with unit rotation at *j*.

The unit moment can also be distributed among the ten strips. Let us identify the moment in each strip. For example, in the 10th strip, the moment is 0.95. Similarly, the moment at other strips can be calculated by proportioning. The values are listed in Table 1.1. Since the moment and the flexural rigidity of the beam is completely known for all the strips, the M/EI diagram for the conjugate beam is developed as shown in Figure 1.20. The M/EI ordinates are also listed in Table 1.1. For example, the M/EI ordinate is given by

M/EI ordinate for 10th strip = $0.95/3.209I = 0.296/EI$

The ends of the beam are marked as *A* and *B*. By taking moment *A*,

$$v_B = \frac{1}{5}\frac{0.5}{EI}\{(0.296\times0.25)+(0.294\times0.75)+(0.288\times1.25)+(0.279\times1.75)$$
$$+(0.265\times2.25)+(0.245\times2.75)+(0.216\times3.25)+(0.176\times3.75)$$
$$+(0.121\times4.25)+(0.046\times4.75)\}$$

Planar Orthogonal Structures

$$v_B = \frac{0.451}{EI} \text{ (downward)}$$

Similarly,

$$v_A = \frac{1}{EI}\{(0.296 + 0.294 + 0.288 + 0.279 + 0.265 + 0.245 + 0.216 + 0.176 + 0.121$$

$$+ 0.0446) \times 0.5 - (0.451)\}$$

$$v_A = \frac{0.662}{EI} \text{ (upward)}$$

Thus, the flexibility matrix of the beam element is given by,

$$[D_r]_i = \frac{1}{EI}\begin{bmatrix} 0.662 & a_{12} \\ -0.451 & a_{22} \end{bmatrix} \qquad (1.34)$$

In order to find the second column of the flexibility matrix, the same procedure as previously mentioned should be followed by applying a unit moment at the kth end, as shown in Figure 1.21. The anticlockwise unit moment is applied at the kth end. It can be seen that the second subscript in the flexibility coefficients is 'k', where the unit rotation is applied. Then, divide the beam into ten strips and calculate the average thickness of the strips. Then, the M/EI ordinate is calculated, as listed in Table 1.2. The M/EI diagram is shown in Figure 1.21. This becomes the loading diagram for the beam.

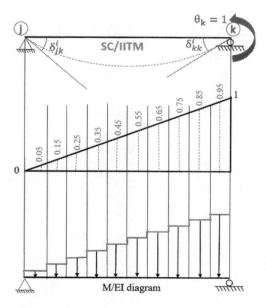

FIGURE 1.21 Simply supported beam with unit rotation at k.

TABLE 1.2
M/EI Ordinates of Strips of the Beam-Unit Rotation at k

Strip	Average Thickness of Strips	Moment of Inertia (I)	Average Moment (M)	M/EI Ordinates
I_1	410=1.025h	1.077I	0.05	0.016/EI
I_2	430=1.075h	1.242I	0.15	0.052/EI
I_3	450=1.125h	1.424I	0.25	0.096/EI
I_4	470=1.175h	1.622I	0.35	0.150/EI
I_5	490=1.225h	1.838I	0.45	0.217/EI
I_6	510=1.275h	2.073I	0.55	0.293/EI
I_7	530=1.325h	2.326I	0.65	0.401/EI
I_8	550=1.375h	2.60I	0.75	0.527/EI
I_9	570=1.425h	2.894I	0.85	0.684/EI
I_{10}	590=1.475h	3.209I	0.95	0.882/EI

To find the vertical reactions, taking the moment about A,

$$v_B = \frac{1}{5}\frac{0.5}{EI}\{(0.016\times 25)+(0.052\times 0.75)+(0.096\times 1.25)+(0.15\times 1.75)$$

$$+(0.217\times 2.25)+(0.293\times 2.75)+(0.401\times 3.25)+(0.527\times 3.75)$$

$$+(0.684\times 4.25)+(0.882\times 4.75)\}$$

$$v_B = \frac{1.210}{EI} \text{ (upward)}$$

$$v_B = \frac{1}{EI}\{(0.016+0.052+0.096+0.15+0.217+0.293+0.401+0.527+0.684$$

$$+0.82)\times 0.5 - 1.210\}$$

$$v_A = \frac{0.45}{EI} \text{ (downward)}$$

Therefore, the flexibility matrix is given by,

$$[D_r]_i = \frac{1}{EI}\begin{bmatrix} 0.662 & -0.450 \\ -0.451 & 1.210 \end{bmatrix}$$

Now, the stiffness matrix is obtained by inverting the flexibility matrix.

$$[K]_i = [D]_i^{-1} = \frac{EI}{0.599}\begin{bmatrix} 1.210 & 0.450 \\ 0.45 & 0.662 \end{bmatrix}$$

Planar Orthogonal Structures

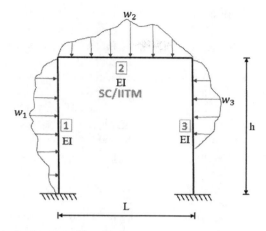

FIGURE 1.22 Planar orthogonal structure.

$$[K]_i = EI \begin{bmatrix} 2.202 & 0.751 \\ 0.750 & 1.105 \end{bmatrix}$$

Consider a beam which is fixed at both the ends and has a varying cross-section. The length of the beam is 5 m. The moment of inertia is already calculated as 1.6×10^9 mm^4. Then, the degrees-of-freedom of the member are marked neglecting the axial deformation, as shown in Figure 1.22. Now, the stiffness matrix can be readily written with the known rotational coefficients. The stiffness matrix is given by,

$$K_i = \begin{bmatrix} 2.02 & 0.751 & 0.554 & -0.554 \\ 0.751 & 1.105 & 0.371 & -0.371 \\ 0.554 & 0.371 & 0.185 & -0.185 \\ -0.554 & -0.371 & -0.185 & 0.185 \end{bmatrix}$$

1.7 PLANAR ORTHOGONAL STRUCTURES

The stiffness method is more generic and easily programmable compared to that of the flexibility method. The stiffness method is also not problem-specific. The stiffness matrix of a fixed beam element can be easily developed with only the rotational coefficients. Let us apply this method to a planar orthogonal structure. Consider a single story single bay frame, as shown in Figure 1.22. Both the ends of the frame are fixed, and the frame is subjected to some arbitrary loading. The height of the frame is taken as 'h' and the flexural rigidity is 'EI'. There are some basic steps to formulate a stiffness method of analysis for solving this problem. The unknowns in stiffness method are displacements which can be translational as well as rotational. Thus, the stiffness method is a generic method that can be applied to any frame under arbitrary loading, because the unknowns in the analysis are not actions but displacements.

It is more or less a well-defined procedure. The members are numbered indexed using square brackets, as shown in Figure 1.22. There are three members in the frame considered for the analysis. Let the joints be numbered in a sequence using circles. There are four joints in the frame.

For every member, it is now important to identify the joint numbering in accordance to the standard fixed beam. Some literature uses the transformation matrix for solving the problem, where the orientation of the member with respect to the origin axis becomes important. But, we will handle this following a different method, without any transformation matrix. So, the elemental member will be a fixed beam irrespective of the boundary conditions of the original beam considered for the analysis.

Step 1: Mark degrees-of-freedom

The elemental member should have both the ends fixed with four degrees-of-freedom in translation and rotation by neglecting axial deformation as shown in Figure 1.23. We will follow the same notations and order for all the problems.

Step 2: Identify unrestrained and restrained displacements at each joint

While numbering the displacements, first label the unrestrained displacements of all the joints. Then, label restrained displacements of all the joints. This is called grouping, which helps in cross-partitioning of the matrix. This will make the analysis more simple, closed form and very easy. For example, consider a frame with both ends fixed as shown in Figure 1.24. The frame has three members and four joints. Joints 2 and 3 are free to rotate and displace, which means the frame can sway. The unrestrained displacements are marked first: θ_1, θ_2 and δ_3, neglecting the axial deformation. Thus, the stiffness matrix in unrestrained degrees-of-freedom is a 3×3 matrix. Joints 1 and 4 are fixed, not allowing any rotation and displacement. The restrained displacements are δ_4, δ_5, θ_6, θ_7, δ_8 and δ_9. Thus, the total number of degrees-of-freedom of the frame is 9. It shows that the size of the stiffness matrix will be 9×9. But, the stiffness matrix of the whole frame can be partitioned.

The stiffness matrix in unrestrained degrees-of-freedom is denoted by k_{uu} and the stiffness matrix in restrained degrees-of-freedom is denoted by k_{rr} as shown in Figure 1.25.

Step 3: To determine the unrestrained displacements

Let us consider the stiffness matrix in term of unrestrained and restrained stiffness matrices. Let the unrestrained and restrained displacements be denoted as Δ_u and Δ_r

FIGURE 1.23 Elemental fixed beam with degrees-of-freedom.

Planar Orthogonal Structures

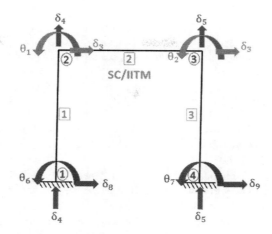

FIGURE 1.24 Single story single bay frame.

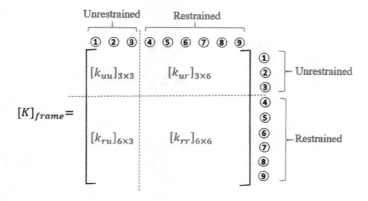

FIGURE 1.25 Partitioning of stiffness matrix.

respectively. The product of the stiffness matrix and the displacement will give the sum of a joint load vector and partial reaction vector, which can again be partitioned into unrestrained and restrained load vectors and reaction vectors respectively. The partitioned stiffness matrix, partitioned displacement vector, partitioned joint load vector and the partitioned reaction vector are given in the following equation:

$$\begin{bmatrix} [K_{uu}] & [K_{ur}] \\ [K_{ru}] & [K_{rr}] \end{bmatrix} \begin{Bmatrix} \Delta_u \\ \Delta_r \end{Bmatrix} = \begin{Bmatrix} J_{Lu} \\ J_{Lr} \end{Bmatrix} + \begin{Bmatrix} [0] \\ R_r \end{Bmatrix} \qquad (1.35)$$

Let us expand the previous equation,

$$[k_{uu}]\{\Delta_u\} = \{J_L\}_u \qquad (1.36)$$

$$\{\Delta_u\} = [k_{uu}]^{-1}\{J_L\}_u \qquad (1.37)$$

Thus, Δ_u can be computed from the previous equation. To compute Δ_u, we need to know joint load vector also.

Step 4: To estimate the joint load vector {J_L}

The joint load vector is derived based on the fixed end moments generated from the applied loads on each member. The fixed end moments for different loading conditions are given in Table 1.3.

[Sign convention: Anticlockwise moments are positive]

The joint load will be simply the reversal of the fixed end moments. End moments due to displacements can also be found out as given in Table 1.4.

Thus, these are the standard formats by which one can find the end moments either caused by the load on the beam or by the displacement on the joint or rotation on the joint. By reversing this, joint loads can be obtained from which the unrestrained displacements can be computed.

1.8 EXAMPLE PROBLEMS

1.8.1 CONTINUOUS BEAM

Analyze the continuous beam and find the reactions and end moments of the beam shown in Figure 1.26:

SOLUTION:

1. *Marking unrestrained and restrained degrees-of-freedom:*
   The unrestrained and restrained degrees-of-freedom are marked on the continuous beam, as shown in Figure 1.27.
   Thus, there are in total six degrees-of-freedom with two unrestrained and four restrained degrees-of-freedom.
   Unrestrained degrees-of-freedom = 2 [θ_1, θ_2]
   Restrained degrees-of-freedom = 4 [θ_3, δ_4, δ_5, δ_6]
   Thus, the unrestrained submatrix will be of size 2×2. The total size of the stiffness matrix will be 6×6. The element stiffness matrix will be 4×4. Let us compare the given problem with a standard fixed beam in Figure 1.28 to obtain the labels. Even though support *B* is simply supported, the basic element is a fixed beam.
   Thus, the labels are

$$AB = [3,1,4,5]$$

$$BC = [1,2,5,6]$$

The labels are ordered in such a manner, the rotations are followed by translations.

Planar Orthogonal Structures

TABLE 1.3
Fixed End Moments

Sl. No.	Beam	M_F^{AB}	M_F^{BA}
1	Fixed beam with central concentrated load: 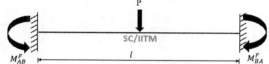	$+\dfrac{Pl}{8}$	$-\dfrac{Pl}{8}$
2	Fixed beam with uniformly distributed load: 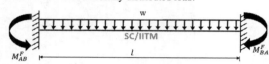	$+\dfrac{wl^2}{12}$	$-\dfrac{wl^2}{12}$
3	Fixed beam with eccentric concentrated load: 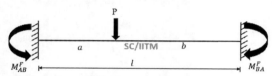	$+\dfrac{pab^2}{l^2}$	$-\dfrac{pba^2}{l^2}$
4	Fixed bean with uniformly varying load: 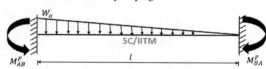	$+\dfrac{W_o l^2}{20}$	$-\dfrac{W_o l^2}{20}$
5	Fixed beam with triangular loading: 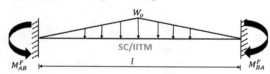	$+\dfrac{5}{96}W_o l^2$	$-\dfrac{5}{96}W_o l^2$
6	Fixed beam with eccentric anticlockwise moment: 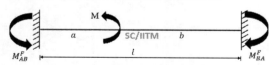	$+\dfrac{M}{l^2}b(b-2a)$	$+\dfrac{M}{l^2}a(2b-a)$
7	Fixed beam with central clockwise moment: 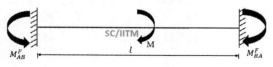	$-\dfrac{M}{4}$	$+\dfrac{M}{4}$

TABLE 1.4
End Moments Due to Displacements

Sl. No.	Beam	M_F^{AB}	M_F^{BA}
1	Beam with rotation at jth end:	$+\dfrac{4EI\theta_A}{l}$	$+\dfrac{2EI\theta_A}{l}$
2	Beam with rotation at kth end:	$+\dfrac{2EI\theta_B}{l}$	$+\dfrac{4EI\theta_B}{l}$
3	Simply supported beam with rotation:	$+\dfrac{3EI\theta_A}{l}$	NIL
4	Beam with settlement of support:	$+\dfrac{6EI\Delta}{l^2}$	$+\dfrac{6EI\Delta}{l^2}$
5	Beam with settlement of support:	$+\dfrac{3EI\Delta}{l^2}$	NIL

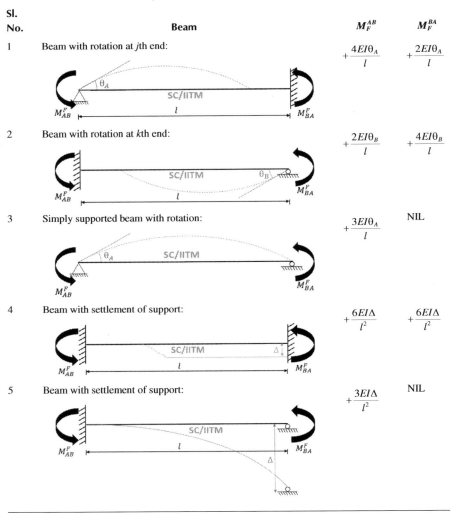

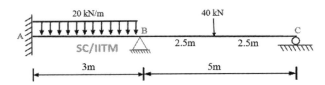

FIGURE 1.26 Continuous beam.

Planar Orthogonal Structures

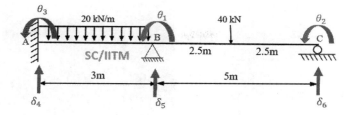

FIGURE 1.27 Degrees-of-freedom.

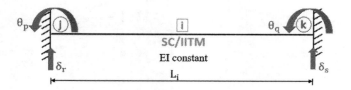

FIGURE 1.28 Standard fixed beam element.

2. *Formulation of stiffness matrix:*
   We already know the formulation of the stiffness matrix for a standard fixed beam element. The stiffness matrix is given by,

$$K_i = \begin{bmatrix} \dfrac{4EI}{l} & \dfrac{2EI}{l} & \dfrac{6EI}{l^2} & -\dfrac{6EI}{l^2} \\ \dfrac{2EI}{l} & \dfrac{4EI}{l} & \dfrac{6EI}{l^2} & -\dfrac{6EI}{l^2} \\ \dfrac{6EI}{l^2} & \dfrac{6EI}{l^2} & \dfrac{12EI}{l^3} & -\dfrac{12EI}{l^3} \\ -\dfrac{6EI}{l^2} & -\dfrac{6EI}{l^2} & -\dfrac{12EI}{l^3} & \dfrac{12EI}{l^3} \end{bmatrix}$$

Thus, the size of the stiffness matrix is 4×4, neglecting axial deformation. For this problem, the element stiffness matrices are given by,

$$K_{AB} = EI \begin{bmatrix} 1.333 & 0.667 & 0.667 & -0.667 \\ 0.667 & 1.333 & 0.667 & -0.667 \\ 0.667 & 0.667 & 0.445 & -0.445 \\ -0.667 & -0.667 & -0.445 & 0.445 \end{bmatrix} \begin{matrix} ③ \\ ① \\ ④ \\ ⑤ \end{matrix}$$

$$\quad\quad\quad ③ \quad ① \quad ④ \quad ⑤$$

$$K_{BC} = EI \begin{bmatrix} \overset{①}{0.80} & \overset{②}{0.40} & \overset{⑤}{0.24} & \overset{⑥}{-0.24} \\ 0.40 & 0.80 & 0.24 & -0.24 \\ 0.24 & 0.24 & 0.096 & -0.096 \\ -0.24 & -0.24 & -0.096 & 0.096 \end{bmatrix} \begin{matrix} ① \\ ② \\ ⑤ \\ ⑥ \end{matrix}$$

Thus, the total stiffness matrix will be written as follows:

$$[K] = \begin{bmatrix} [k_{uu}]_{2\times 2} & [k_{ur}]_{2\times 4} \\ [k_{ru}]_{4\times 2} & [k_{rr}]_{4\times 4} \end{bmatrix}$$

We need only the stiffness matrix of unrestrained degrees-of-freedom.

$$K_{UU} = EI \begin{bmatrix} \overset{①}{2.133} & \overset{②}{0.4} \\ 0.4 & 0.8 \end{bmatrix} \begin{matrix} ① \\ ② \end{matrix}$$

Then

$$K_{UU}^{-1} = \frac{1}{1.546EI} \begin{bmatrix} 2.133 & 0.4 \\ 0.4 & 0.8 \end{bmatrix}$$

3. *Calculation of fixed end moments:*

Now, the fixed end moments are calculated to find the unrestrained displacements. The fixed end moments and reactions of the members are shown in Figure 1.29.

$$M_{AB}^F = \frac{wl^2}{12} = \frac{20 \times 3^2}{12} = +15 \text{ kNm}$$

$$M_{BA}^F = -\frac{wl^2}{12} = -\frac{20 \times 3^2}{12} = -15 \text{ kNm}$$

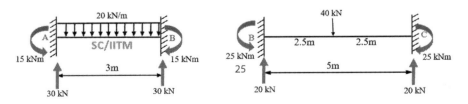

FIGURE 1.29 Fixed end moments.

Planar Orthogonal Structures

$$V_A = +30 \text{ kN}$$

$$V_B = +30 \text{ kN}$$

$$M_{BC}^F = \frac{pl}{8} = \frac{40 \times 5}{8} = +25 \text{ kNm}$$

$$M_{CB}^F = -\frac{pl}{8} = -\frac{40 \times 5}{8} = -25 \text{ kNm}$$

$$V_B = +20 \text{ kN}$$

$$V_C = +20 \text{ kN}$$

Thus, for the whole beam, the fixed end moments are shown in Figure 1.30.

4. *Calculation of joint load vectors:*
The joint loads will be the reversal of the fixed end moments and reactions. The joint load for the continuous beam is shown in Figure 1.31.
Now, the joint load vector is given by,

$$J_L = \left\{ \begin{array}{c} J_{LU} \\ --- \\ J_{LR} \end{array} \right\}$$

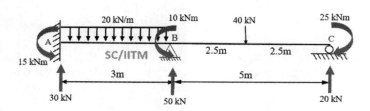

FIGURE 1.30 Fixed end moments for the continuous beam.

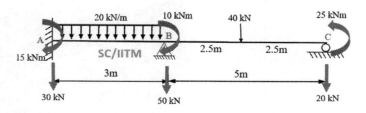

FIGURE 1.31 Joint load vector for the continuous beam.

$$J_L = \begin{Bmatrix} -10 \\ +25 \\ -15 \\ -30 \\ -50 \\ -20 \end{Bmatrix} \begin{matrix} \text{①} \\ \text{②} \\ \text{③} \\ \text{④} \\ \text{⑤} \\ \text{⑥} \end{matrix}$$

The unrestrained joint load vector is given by,

$$J_{LU} = \begin{Bmatrix} -10 \\ +25 \end{Bmatrix} \begin{matrix} \text{①} \\ \text{②} \end{matrix}$$

5. *Calculation of displacements:*
Now,

$$\{\Delta_u\} = [k_{uu}]^{-1} \{J_L\}_u$$

$$\begin{Bmatrix} \theta_1 \\ \theta_2 \end{Bmatrix} = \frac{1}{1.546EI} \begin{bmatrix} 0.8 & -0.4 \\ -0.4 & 2.133 \end{bmatrix} \begin{Bmatrix} -10 \\ +25 \end{Bmatrix} = \begin{Bmatrix} -18/1.546EI \\ 57.325/1.546EI \end{Bmatrix}$$

The vector of restrained displacements will be zero.

$$\begin{Bmatrix} \theta_3 \\ \delta_4 \\ \delta_5 \\ \delta_6 \end{Bmatrix} = 0$$

6. *Calculation of end moment and shear:*
The general equation to find the end moment of the *i*th beam is,

$$[M]_i = k_i \delta_i + (\text{FEM})_i$$

The equation can be rewritten element-wise as follows:

$$M_{AB} = [k]_{AB} \{\delta\} + \{\text{FEM}\}$$

$$\begin{Bmatrix} M_3 \\ M_1 \\ V_4 \\ V_5 \end{Bmatrix} = EI \begin{bmatrix} 1.333 & 0.667 & 0.667 & -0.667 \\ 0.667 & 1.333 & 0.667 & -0.667 \\ 0.667 & 0.667 & 0.445 & -0.445 \\ -0.667 & -0.667 & -0.445 & 0.445 \end{bmatrix} \begin{Bmatrix} \theta_3 \\ \theta_1 \\ \delta_4 \\ \delta_5 \end{Bmatrix} + \begin{Bmatrix} -15 \\ -15 \\ +30 \\ +30 \end{Bmatrix} = \begin{Bmatrix} 7.234 \\ -30.52 \\ 22.234 \\ 37.766 \end{Bmatrix}$$

Planar Orthogonal Structures

$$M_{BC} = [k]_{BC} \{\delta\} + \{FEM\}$$

$$\begin{Bmatrix} M_1 \\ M_2 \\ V_5 \\ V_6 \end{Bmatrix} = EI \begin{bmatrix} 0.8 & 0.4 & 0.24 & -0.24 \\ 0.4 & 0.8 & 0.24 & -0.24 \\ 0.24 & 0.24 & 0.096 & -0.096 \\ -0.24 & -0.24 & -0.096 & 0.096 \end{bmatrix} \begin{Bmatrix} \theta_1 \\ \theta_2 \\ \delta_5 \\ \delta_6 \end{Bmatrix} + \begin{Bmatrix} 25 \\ -25 \\ 20 \\ 20 \end{Bmatrix} = \begin{Bmatrix} 30.517 \\ 0 \\ 26.105 \\ 13.895 \end{Bmatrix}$$

The member end moments and shear are shown in Figure 1.32.

Check for span AB:

$$\sum V = 0,$$

$$22.234 + 37.766 = 20 \times 3 = 60 \text{ kN}$$

$$M_A = (20 \times 3 \times 1.5) + 30.52 - 37.766(3) = 7.234 \text{ kNm}$$

Hence, Ok.

Check for span BC:

$$\sum V = 0,$$

$$(26.105 + 13.895) = 40 \text{ kN}$$

$$M_B = (40 \times 2.5) - (13.895 \times 5) = 30.52 \text{ kNm}$$

Hence, Ok.

By superimposing the end moments of the members into the continuous beam, the final end moments and shear are obtained, as shown in Figure 1.33.

Check:
We can say that,

$$[k_{ru}]\{\Delta_u\} - \{J_l\}_r = \{R_r\}$$

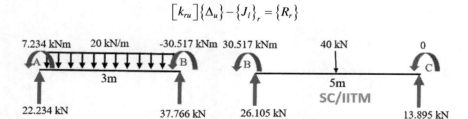

FIGURE 1.32 Member end moments and shear.

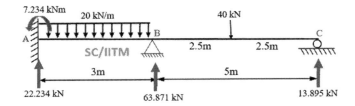

FIGURE 1.33 Final end moments and shear.

$$[k_{ru}] = EI \begin{bmatrix} 0.667 & 0 \\ 0.667 & 0 \\ -0.427 & 0.24 \\ -0.24 & -0.24 \end{bmatrix}$$

$$\{\Delta_u\} \frac{1}{1.546EI} \begin{Bmatrix} -18 \\ 57.325 \end{Bmatrix}$$

$$\{J_l\}_r = \begin{Bmatrix} -15 \\ -30 \\ -50 \\ -20 \end{Bmatrix} \begin{matrix} ③ \\ ④ \\ ⑤ \\ ⑥ \end{matrix}$$

Thus,

$$\{R_r\} = \begin{Bmatrix} M_3 \\ V_4 \\ V_5 \\ V_6 \end{Bmatrix} = \begin{Bmatrix} 7.234 \\ 22.234 \\ 63.81 \\ 13.895 \end{Bmatrix}.$$

Hence, Ok.

1.8.2 Computer Program for Continuous Beam

```
%% stiffness matrix method
% Input
clc;
clear;
n = 2; % number of members
I = [1 1]; %Moment of inertis in m4
L = [3 5]; % length in m
uu = 2; % Number of unrestrained degrees of freedom
ur = 4; % Number of restrained degrees of freedom
uul = [1 2]; % global labels of unrestrained dof
url = [3 4 5 6]; % global labels of restrained dof
l1 = [3 1 4 5]; % Global labels for member 1
l2 = [1 2 5 6]; % Global labels for member 2
```

Planar Orthogonal Structures 39

```
l= [l1; l2];
dof = uu+ur;
Ktotal = zeros (dof);
fem1= [15 -15 30 30]; % Local Fixed end moments of member 1
fem2= [25 -25 20 20]; % Local Fixed end moments of member 2

%% Creation of joint load vector
jl= [-10; 25; -15; -30; -50; -20]; % values given in kN or kNm
jlu = [-10; 25]; % load vector in unrestrained dof

%% rotation coefficients for each member
rc1 = 4.*I./L;
rc2 = 2.*I./L;

%% stiffness matrix 4 by 4 (axial deformation neglected)
for i = 1:n
    Knew = zeros (dof);
    k1 = [rc1(i); rc2(i); (rc1(i)+rc2(i))/L(i);
(-(rc1(i)+rc2(i))/L(i))];
    k2 = [rc2(i); rc1(i); (rc1(i)+rc2(i))/L(i);
(-(rc1(i)+rc2(i))/L(i))];
    k3 = [(rc1(i)+rc2(i))/L(i); (rc1(i)+rc2(i))/L(i);
(2*(rc1(i)+rc2(i))/(L(i)^2)); (-2*(rc1(i)+rc2(i))/(L(i)^2))];
    k4 = -k3;
    K = [k1 k2 k3 k4];
    fprintf ('Member Number =');
    disp (i);
    fprintf ('Local Stiffness matrix of member, [K] = \n');
    disp (K);
    for p = 1:4
        for q = 1:4
            Knew((l(i,p)),(l(i,q))) =K(p,q);
        end
    end
    Ktotal = Ktotal + Knew;
    if i == 1
        Kg1=K;
    else
        Kg2 = K;
    end
end
fprintf ('Stiffness Matrix of complete structure, [Ktotal] = \n');
disp (Ktotal);
Kunr = zeros(uu);
for x=1:uu
    for y=1:uu
        Kunr(x,y)= Ktotal(x,y);
    end
end
fprintf ('Unrestrained Stiffness sub-matrix, [Kuu] = \n');
disp (Kunr);
```

```matlab
KuuInv= inv(Kunr);
fprintf ('Inverse of Unrestrained Stiffness sub-matrix,
[KuuInverse] = \n');
disp (KuuInv);

%% Calculation of displacements
delu = KuuInv*jlu;
fprintf ('Joint Load vector, [Jl] = \n');
disp (jl');
fprintf ('Unrestrained displacements, [DelU] = \n');
disp (delu');
delr = zeros (ur,1);
del = [delu; delr];
deli= zeros (4,1);
for i = 1:n
    for p = 1:4
        deli(p,1) = del((l(i,p)),1) ;
    end
    if i == 1
            delbar1 = deli;
            mbar1= (Kg1 * delbar1)+fem1';
            fprintf ('Member Number =');
            disp (i);
            fprintf ('Global displacement matrix [DeltaBar] =
            \n');
            disp (delbar1');
            fprintf ('Global End moment matrix [MBar] = \n');
            disp (mbar1');
        else
            delbar2 = deli;
            mbar2= (Kg2 * delbar2)+fem2';
            fprintf ('Member Number =');
            disp (i);
            fprintf ('Global displacement matrix [DeltaBar] =
            \n');
            disp (delbar2');
            fprintf ('Global End moment matrix [MBar] = \n');
            disp (mbar2');
    end
end
%% check
mbar = [mbar1'; mbar2'];
jf = zeros(dof,1);
for a=1:n
    for b=1:4 % size of k matrix
        d = l(a,b);
        jfnew = zeros(dof,1);
        jfnew(d,1)=mbar(a,b);
        jf=jf+jfnew;
    end
end
```

Planar Orthogonal Structures

```
fprintf ('Joint forces = \n');
disp (jf');
```

MATLAB® output:

```
Member Number = 1
Local Stiffness matrix of member, [K] =

    1.3333    0.6667    0.6667   -0.6667
    0.6667    1.3333    0.6667   -0.6667
    0.6667    0.6667    0.4444   -0.4444
   -0.6667   -0.6667   -0.4444    0.4444

Member Number = 2
Local Stiffness matrix of member, [K] =

    0.8000    0.4000    0.2400   -0.2400
    0.4000    0.8000    0.2400   -0.2400
    0.2400    0.2400    0.0960   -0.0960
   -0.2400   -0.2400   -0.0960    0.0960

Stiffness Matrix of complete structure, [Ktotal] =

    2.1333    0.4000    0.6667    0.6667   -0.4267   -0.2400
    0.4000    0.8000         0         0    0.2400   -0.2400
    0.6667         0    1.3333    0.6667   -0.6667         0
    0.6667         0    0.6667    0.4444   -0.4444         0
   -0.4267    0.2400   -0.6667   -0.4444    0.5404   -0.0960
   -0.2400   -0.2400         0         0   -0.0960    0.0960

Unrestrained Stiffness sub-matrix, [Kuu] =

    2.1333    0.4000
    0.4000    0.8000

Inverse of Unrestrained Stiffness sub-matrix, [KuuInverse] =

    0.5172   -0.2586
   -0.2586    1.3793

Joint Load vector, [Jl] =

   -10    25   -15   -30   -50   -20

Unrestrained displacements, [DelU] =

  -11.6379   37.0690

Member Number = 1
Global displacement matrix [DeltaBar] =

     0  -11.6379    0    0
```

```
Global End moment matrix [MBar] =
   7.2414   -30.5172   22.2414   37.7586

Member Number = 2
Global displacement matrix [DeltaBar] =
  -11.6379   37.0690   0   0

Global End moment matrix [MBar] =
   30.5172   -0.0000   26.1034   13.8966

Joint forces =
   0   -0.0000   7.2414   22.2414   63.8621   13.8966
```

1.8.3 Orthogonal Frame

Analyze the single story single bay frame shown in Figure 1.34:

1. *Marking unrestrained and restrained degrees-of-freedom:*

The unrestrained and restrained degrees-of-freedom are marked on the continuous beam, as shown in Figure 1.35.
Thus, the size of the stiffness matrix will be 9×9.
Unrestrained degrees-of-freedom = 2 [θ_1, θ_2]
Restrained degrees-of-freedom = 7 [$\theta_3, \theta_4, \delta_4, \delta_5, \delta_6, \delta_8, \delta_9$]
Thus, the unrestrained submatrix size will be 2×2. The total size of the stiffness matrix will be 9×9. The element stiffness matrix will be 4×4, neglecting axial deformation.

$$[K] = \begin{bmatrix} [k_{uu}]_{2\times 2} & [k_{ur}]_{2\times 7} \\ [k_{ru}]_{7\times 2} & [k_{rr}]_{7\times 7} \end{bmatrix}$$

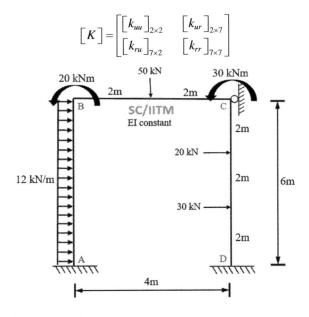

FIGURE 1.34 Frame example.

Planar Orthogonal Structures

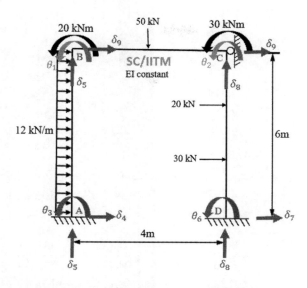

FIGURE 1.35 Unrestrained and restrained degrees-of-freedom.

The orientation and the labels of the members are given in the following table:

Member	jth End	kth End	Degrees-of-Freedom Labels
BA	B	A	1, 3, 9, 4
BC	B	C	1, 2, 5, 8
CD	C	D	2, 6, 9, 7

Thus,

$$\{\Delta\}_{9\times 1} = \begin{Bmatrix} \theta_1 \\ \theta_2 \\ \theta_3 \\ \delta_4 \\ \delta_5 \\ \delta_6 \\ \delta_7 \\ \delta_8 \\ \delta_9 \end{Bmatrix} = \begin{Bmatrix} [\Delta_u]_{2\times 1} \\ [\Delta_r]_{7\times 1} \end{Bmatrix}$$

2. *Formulation of stiffness matrix:*

The stiffness matrix is given by,

$$K_i = \begin{bmatrix} \frac{4EI}{l} & \frac{2EI}{l} & \frac{6EI}{l^2} & -\frac{6EI}{l^2} \\ \frac{2EI}{l} & \frac{4EI}{l} & \frac{6EI}{l^2} & -\frac{6EI}{l^2} \\ \frac{6EI}{l^2} & \frac{6EI}{l^2} & \frac{12EI}{l^3} & -\frac{12EI}{l^3} \\ -\frac{6EI}{l^2} & -\frac{6EI}{l^2} & -\frac{12EI}{l^3} & \frac{12EI}{l^3} \end{bmatrix}$$

Thus, the size of the stiffness matrix is 4×4, neglecting axial deformation. For this problem, the element stiffness matrices are given by,

$$\begin{array}{cccc} \textcircled{1} & \textcircled{3} & \textcircled{9} & \textcircled{4} \end{array}$$

$$K_{AB} = EI \begin{bmatrix} 1.333 & 0.667 & 0.333 & -0.333 \\ 0.667 & 1.333 & 0.333 & -0.333 \\ 0.333 & 0.333 & 0.111 & -0.111 \\ -0.333 & -0.333 & -0.111 & 0.111 \end{bmatrix} \begin{array}{c} \textcircled{1} \\ \textcircled{3} \\ \textcircled{9} \\ \textcircled{4} \end{array}$$

$$\begin{array}{cccc} \textcircled{1} & \textcircled{2} & \textcircled{5} & \textcircled{8} \end{array}$$

$$K_{BC} = EI \begin{bmatrix} 1 & 0.5 & 0.375 & -0.375 \\ 0.5 & 1 & 0.375 & -0.375 \\ 0.375 & 0.375 & 0.188 & -0.188 \\ -0.375 & -0.375 & -0.188 & 0.188 \end{bmatrix} \begin{array}{c} \bigcirc \\ \textcircled{2} \\ \textcircled{5} \\ \textcircled{8} \end{array}$$

$$\begin{array}{cccc} \textcircled{2} & \textcircled{6} & \textcircled{9} & \textcircled{7} \end{array}$$

$$K_{CD} = EI \begin{bmatrix} 1.33 & 0.667 & 0.333 & -0.333 \\ 0.667 & 1.333 & 0.333 & -0.333 \\ 0.333 & 0.333 & 0.111 & -0.111 \\ -0.333 & -0.333 & -0.111 & 0.111 \end{bmatrix} \begin{array}{c} \textcircled{2} \\ \textcircled{6} \\ \textcircled{9} \\ \textcircled{7} \end{array}$$

We need only the stiffness matrix of unrestrained degrees-of-freedom.

$$\begin{array}{cc} \textcircled{1} & \textcircled{2} \end{array}$$

$$K_{UU} = EI \begin{bmatrix} 2.333 & 0.5 \\ 0.5 & 2.333 \end{bmatrix} \begin{array}{c} \textcircled{1} \\ \textcircled{2} \end{array}$$

Then

$$K_{UU}^{-1} = \frac{1}{5.193EI} \begin{bmatrix} 2.333 & -0.5 \\ -0.5 & 2.333 \end{bmatrix}$$

Planar Orthogonal Structures

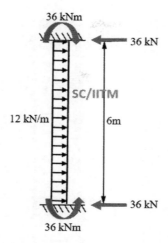

FIGURE 1.36 Fixed end moments for span AB.

3. *Calculation of fixed end moments:*
The fixed end moments and reactions of the members are given subsequently:
For member AB (Figure 1.36),

$$M_{AB}^F = -\frac{wl^2}{12} = -\frac{12 \times 6^2}{12} = -36 \text{ kNm}$$

$$M_{BA}^F = +36 \text{ kNm}$$

$$V_A = -36 \text{ kN}$$

$$V_B = -36 \text{ kN}$$

For member BC (Figure 1.37),

$$M_{BC}^F = \frac{pl}{8} = \frac{50 \times 4}{8} = +25 \text{ kNm}$$

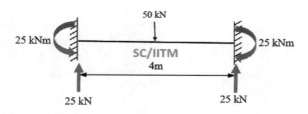

FIGURE 1.37 Fixed end moments for span BC.

$$M_{CB}^F = -25 \text{ kNm}$$

$$V_B = +25 \text{ kN}$$

$$V_C = +25 \text{ kN}$$

For member CD (Figure 1.38),
For 20 kN load,

$$M_{CD}^F = \frac{pab^2}{l^2} = \frac{20 \times 2 \times 4^2}{6^2} = +17.778 \text{ kNm}$$

$$M_{DC}^F = \frac{pa^2b}{l^2} = \frac{20 \times 2^2 \times 4}{6^2} = -8.889 \text{ kNm}$$

$$V_C = +13.333 \text{ kN}$$

$$V_D = +6.667 \text{ kN}$$

For 30 kN load,

$$M_{CD}^F = \frac{pab^2}{l^2} = \frac{30 \times 4 \times 2^2}{6^2} = +13.333 \text{ kNm}$$

$$M_{DC}^F = \frac{pa^2b}{l^2} = \frac{30 \times 2 \times 4^2}{6^2} = -26.667 \text{ kNm}$$

$$V_C = +10 \text{ kN}$$

$$V_D = +20 \text{ kN}$$

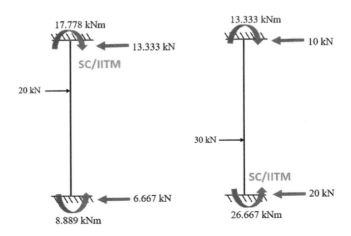

FIGURE 1.38 Fixed end moments for span BC.

Planar Orthogonal Structures

Thus, for the member CD, Fixed end moments and reactions are given by,

$$M_{CD}^F = -17.778 - 13.333 = -31.111 \text{ kNm}$$

$$M_{DC}^F = +8.889 + 26.667 = +35.556 \text{ kNm}$$

$$V_C = -13.333 - 10 = -23.333 \text{ kN}$$

$$V_D = -6.667 - 20 = -26.667 \text{ kN}$$

Thus, for the whole beam, the fixed end moments are written as follows:

$$AB = \begin{Bmatrix} -36 \\ +36 \\ -36 \\ -36 \end{Bmatrix}, \quad BC = \begin{Bmatrix} +25 \\ -25 \\ +25 \\ +25 \end{Bmatrix}, \quad CD = \begin{Bmatrix} -31.111 \\ +35.556 \\ -23.333 \\ -26.667 \end{Bmatrix}$$

4. Calculation of joint load vectors:

The size of the joint load vector is 9×1.

$$\{J_L\}_{9\times1} = \begin{Bmatrix} +31 \\ 86.111 \\ -36 \\ +36 \\ -25 \\ -35.556 \\ +26.667 \\ -25.000 \\ +59.333 \end{Bmatrix} \begin{matrix} ① \\ ② \\ ③ \\ ④ \\ ⑤ \\ ⑥ \\ ⑦ \\ ⑧ \\ ⑨ \end{matrix}$$

The joint load vector can be split into unrestrained and restrained degrees-of-freedom.

$$\{J_L\} = \begin{Bmatrix} (J_L)_u \\ _{2\times1} \\ --- \\ (J_L)_r \\ _{7\times1} \end{Bmatrix}_{9\times1}$$

Thus, $J_{Lu} = \begin{Bmatrix} +31 \\ 86.111 \end{Bmatrix} \begin{matrix} ① \\ ② \end{matrix}$

5. Calculation of displacements:

Now,

$$\{\Delta_u\} = [k_{uu}]^{-1}\{J_L\}_u$$

$$\begin{Bmatrix} \theta_1 \\ \theta_2 \end{Bmatrix} = \frac{1}{5.193EI} \begin{bmatrix} 2.333 & -0.5 \\ -0.5 & 2.333 \end{bmatrix} \begin{Bmatrix} 31 \\ 86.111 \end{Bmatrix} = \frac{1}{5.193EI} \begin{Bmatrix} 29.268 \\ 185.397 \end{Bmatrix} \text{ radians}$$

6. Calculation of end moment and shear:

The general equation to find the end moment of the ith beam is,

$$[M]_i = k_i \delta_i + (\text{FEM})_i$$

The equation can be rewritten element-wise as follows:

$$M_{AB} = [k]_{AB}\{\delta\} + \{\text{FEM}\}$$

$$\begin{Bmatrix} M_1 \\ M_3 \\ V_9 \\ V_4 \end{Bmatrix} = EI \begin{bmatrix} 1.333 & 0.667 & 0.333 & -0.333 \\ 0.667 & 1.333 & 0.333 & -0.333 \\ 0.333 & 0.333 & 0.111 & -0.111 \\ -0.333 & -0.333 & -0.111 & 0.111 \end{bmatrix} \begin{Bmatrix} \theta_1 \\ \theta_3 \\ \delta_9 \\ \delta_4 \end{Bmatrix} + \begin{Bmatrix} -36 \\ +36 \\ -36 \\ +36 \end{Bmatrix}$$

We know that, $\theta_1 = \dfrac{29.268}{5.193EI}$, $\theta_3 = \delta_9 = \delta_4 = 0$.

Thus, $\begin{Bmatrix} M_1 \\ M_3 \\ V_9 \\ V_4 \end{Bmatrix} = \begin{Bmatrix} -28.487 \\ 39.759 \\ 34.123 \\ -34.123 \end{Bmatrix}$

$$M_{BC} = [k]_{BC}\{\delta\} + \{\text{FEM}\}$$

$$\begin{Bmatrix} M_1 \\ M_2 \\ V_5 \\ V_8 \end{Bmatrix} = EI \begin{bmatrix} 1 & 0.5 & 0.375 & -0.375 \\ 0.5 & 1 & 0.375 & -0.375 \\ 0.375 & 0.375 & 0.188 & -0.188 \\ -0.375 & -0.375 & -0.188 & 0.188 \end{bmatrix} \begin{Bmatrix} \theta_1 \\ \theta_2 \\ \delta_5 \\ \delta_8 \end{Bmatrix} + \begin{Bmatrix} 25 \\ -25 \\ 25 \\ 25 \end{Bmatrix} = \begin{Bmatrix} 48.487 \\ 13.519 \\ 40.502 \\ 9.498 \end{Bmatrix}$$

Similarly, for the member CD,

$$M_{CD} = [k]_{CD}\{\delta\} + \{\text{FEM}\}$$

Planar Orthogonal Structures

$$\begin{Bmatrix} M_2 \\ M_6 \\ V_9 \\ V_7 \end{Bmatrix} = EI \begin{bmatrix} 1.333 & 0.667 & 0.333 & -0.333 \\ 0.667 & 1.333 & 0.333 & -0.333 \\ 0.333 & 0.333 & 0.111 & -0.111 \\ -0.333 & -0.333 & -0.111 & 0.111 \end{bmatrix} \begin{Bmatrix} \theta_2 \\ \theta_6 \\ \delta_9 \\ \delta_7 \end{Bmatrix} + \begin{Bmatrix} -31.111 \\ 35.556 \\ -23.333 \\ -26.667 \end{Bmatrix} = \begin{Bmatrix} 16.479 \\ 59.369 \\ -11.444 \\ -38.556 \end{Bmatrix}$$

The member end moments and shear are shown in Figure 1.39, and the final moments in Figure 1.40.

Check:
In member AB,

$$\sum F_H = 0$$

$$(12 \times 6) - 34.123 - 37.877 = 0$$

$$\sum M_A = 0$$

$$+28.487 + (12 \times 6 \times 3) - (34.123 \times 6) + 39.75 = 0$$

In member BC,

$$\sum F_V = 0$$

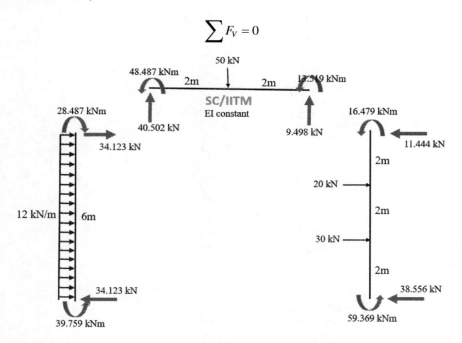

FIGURE 1.39 Member fixed end moments and shear.

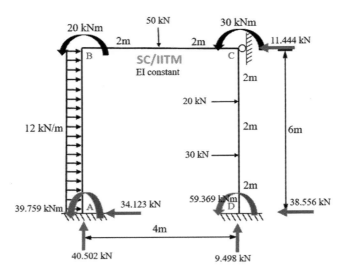

FIGURE 1.40 Final end moments and shear.

$$40.502 + 9.798 - 50 = 0$$

$$\sum M_B = 0$$

$$-13.159 - (9.498 \times 4) + (50 \times 2) + 98.489 = 0$$

In member CD,

$$\sum F_H = 0$$

$$20 + 30 - 11.444 - 38.556 = 0$$

$$\sum M_D = 0$$

$$+16.479 + (11.444 \times 6) - (20 \times 4) - (30 \times 2) + 59.369 = 0$$

1.8.4 Computer Program for Orthogonal Frame

```
%% stiffness matrix method
% Input
clc;
clear;
n = 3; % number of members
I = [2 1 2]; %Moment of inertis in m4
L = [6 4 6]; % length in m
uu = 2; % Number of unrestrained degrees of freedom
```

Planar Orthogonal Structures

```
ur = 7; % Number of restrained degrees of freedom
uul = [1 2]; % global labels of unrestrained dof
url = [3 4 5 6 7 8 9]; % global labels of restrained dof
l1 = [1 3 9 4]; % Global labels for member 1
l2 = [1 2 5 8]; % Global labels for member 2
l3 = [2 6 9 7]; % Global labels for member 3
l= [l1; l2; l3];
dof = uu+ur;
Ktotal = zeros (dof);
fem1= [-36 36 -36 -36]; % Local Fixed end moments of member 1
fem2= [25 -25 25 25]; % Local Fixed end moments of member 2
fem3= [-31.111 35.556 -23.333 -26.667]; % Local Fixed end
moments of member 3

%% rotation coefficients for each member
rc1 = 4.*I./L;
rc2 = 2.*I./L;

%% stiffness matrix 4 by 4 (axial deformation neglected)
for i = 1:n
    Knew = zeros (dof);
    k1 = [rc1(i); rc2(i); (rc1(i)+rc2(i))/L(i);
(-(rc1(i)+rc2(i))/L(i))];
    k2 = [rc2(i); rc1(i); (rc1(i)+rc2(i))/L(i);
(-(rc1(i)+rc2(i))/L(i))];
    k3 = [(rc1(i)+rc2(i))/L(i); (rc1(i)+rc2(i))/L(i);
(2*(rc1(i)+rc2(i))/(L(i)^2)); (-2*(rc1(i)+rc2(i))/(L(i)^2))];
    k4 = -k3;
    K = [k1 k2 k3 k4];
    fprintf ('Member Number =');
    disp (i);
    fprintf ('Local Stiffness matrix of member, [K] = \n');
    disp (K);
    for p = 1:4
        for q = 1:4
            Knew((l(i,p)),(l(i,q))) =K(p,q);
        end
    end
    Ktotal = Ktotal + Knew;
    if i == 1
        Kg1=K;
    elseif i == 2
        Kg2 = K;
    else
        Kg3=K;
    end
end
fprintf ('Stiffness Matrix of complete structure, [Ktotal] = \n');
disp (Ktotal);
Kunr = zeros(uu);
for x=1:uu
```

```matlab
        for y=1:uu
            Kunr(x,y)= Ktotal(x,y);
        end
end
fprintf ('Unrestrained Stiffness sub-matrix, [Kuu] = \n');
disp (Kunr);
KuuInv= inv(Kunr);
fprintf ('Inverse of Unrestrained Stiffness sub-matrix,
[KuuInverse] = \n');
disp (KuuInv);

%% Creation of joint load vector
jl= [31; 86.111; -36; 36; -25; -35.556; 26.665; -25; 59.333];
% values given in kN or kNm
jlu = [31; 86.111]; % load vector in unrestrained dof
delu = KuuInv*jlu;
fprintf ('Joint Load vector, [Jl] = \n');
disp (jl');
fprintf ('Unrestrained displacements, [DelU] = \n');
disp (delu');
delr = zeros (ur,1);
del = [delu; delr];
deli= zeros (4,1);
for i = 1:n
    for p = 1:4
        deli(p,1) = del((l(i,p)),1) ;
    end
    if i == 1
            delbar1 = deli;
            mbar1= (Kg1 * delbar1)+fem1';
            fprintf ('Member Number =');
            disp (i);
            fprintf ('Global displacement matrix [DeltaBar] = \n');
            disp (delbar1');
            fprintf ('Global End moment matrix [MBar] = \n');
            disp (mbar1');
        elseif i == 2
            delbar2 = deli;
            mbar2= (Kg2 * delbar2)+fem2';
            fprintf ('Member Number =');
            disp (i);
            fprintf ('Global displacement matrix [DeltaBar] = \n');
            disp (delbar2');
            fprintf ('Global End moment matrix [MBar] = \n');
            disp (mbar2');
        else
            delbar3 = deli;
            mbar3= (Kg3 * delbar3)+fem3';
            fprintf ('Member Number =');
            disp (i);
            fprintf ('Global displacement matrix [DeltaBar] = \n');
```

Planar Orthogonal Structures

```
                disp (delbar3');
                fprintf ('Global End moment matrix [MBar] = \n');
                disp (mbar3');
        end
end
%% check
mbar = [mbar1'; mbar2'; mbar3'];
jf = zeros(dof,1);
for a=1:n
    for b=1:4 % size of k matrix
        d = l(a,b);
        jfnew = zeros(dof,1);
        jfnew(d,1)=mbar(a,b);
        jf=jf+jfnew;
    end
end
fprintf ('Joint forces = \n');
disp (jf');
```

MATLAB output:

```
Member Number =  1
Local Stiffness matrix of member, [K] =

    1.3333    0.6667    0.3333   -0.3333
    0.6667    1.3333    0.3333   -0.3333
    0.3333    0.3333    0.1111   -0.1111
   -0.3333   -0.3333   -0.1111    0.1111

Member Number =  2
Local Stiffness matrix of member, [K] =

    1.0000    0.5000    0.3750   -0.3750
    0.5000    1.0000    0.3750   -0.3750
    0.3750    0.3750    0.1875   -0.1875
   -0.3750   -0.3750   -0.1875    0.1875

Member Number =  3
Local Stiffness matrix of member, [K] =

    1.3333    0.6667    0.3333   -0.3333
    0.6667    1.3333    0.3333   -0.3333
    0.3333    0.3333    0.1111   -0.1111
   -0.3333   -0.3333   -0.1111    0.1111

Stiffness Matrix of complete structure, [Ktotal] =

    2.3333  0.5000  0.6667 -0.3333  0.3750       0        0 -0.3750  0.3333
    0.5000  2.3333       0       0  0.3750  0.6667  -0.3333 -0.3750  0.3333
    0.6667       0  1.3333 -0.3333       0       0        0       0  0.3333
   -0.3333       0 -0.3333  0.1111       0       0        0       0 -0.1111
    0.3750  0.3750       0       0  0.1875       0        0 -0.1875       0
```

```
            0    0.6667         0         0         0   1.3333  -0.3333         0   0.3333
            0   -0.3333         0         0         0  -0.3333   0.1111         0  -0.1111
      -0.3750  -0.3750         0         0  -0.1875         0         0   0.1875         0
       0.3333   0.3333   0.3333  -0.1111         0   0.3333  -0.1111         0   0.2222
```

Unrestrained Stiffness sub-matrix, [Kuu] =

```
    2.3333    0.5000
    0.5000    2.3333
```

Inverse of Unrestrained Stiffness sub-matrix, [KuuInverse] =

```
    0.4492   -0.0963
   -0.0963    0.4492
```

Joint Load vector, [Jl] =

```
    31.0000  86.1110  -36.0000  36.0000  -25.0000  -35.5560  26.6650  -25.0000  59.3330
```

Unrestrained displacements, [DelU] =

```
    5.6364    35.6969
```

Member Number =   1
Global displacement matrix [DeltaBar] =

```
    5.6364    0    0    0
```

Global End moment matrix [MBar] =

```
   -28.4848    39.7576    -34.1212    -37.8788
```

Member Number =   2
Global displacement matrix [DeltaBar] =

```
    5.6364    35.6969    0    0
```

Global End moment matrix [MBar] =

```
    48.4848    13.5151    40.5000    9.5000
```

Member Number =   3
Global displacement matrix [DeltaBar] =

```
    35.6969    0    0    0
```

Global End moment matrix [MBar] =

```
    16.4849    59.3539    -11.4340    -38.5660
```

Joint forces =

```
    20.0000    30.0000    39.7576   -37.8788
    40.5000    59.3539   -38.5660    9.5000   -45.5552
```

Planar Orthogonal Structures

1.8.5 Step Frame

Analyze the frame shown in Figure 1.41:

1. *Marking unrestrained and restrained degrees-of-freedom:*

The unrestrained and restrained degrees-of-freedom are marked on the continuous beam as shown in Figure 1.42.

The total number of degrees-of-freedom is 13. Thus, the size of the stiffness matrix will be 13×13.

Unrestrained degrees-of-freedom = 7 [$\theta_1, \theta_2, \theta_3, \theta_4, \delta_5, \delta_6, \delta_7$]
Restrained degrees-of-freedom = 6 [$\theta_8, \theta_{11}, \delta_9, \delta_{10}, \delta_{12}, \delta_{13}$]

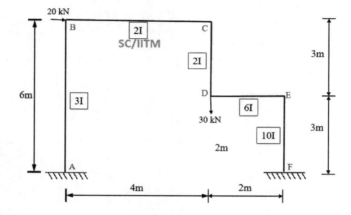

FIGURE 1.41 Frame example.

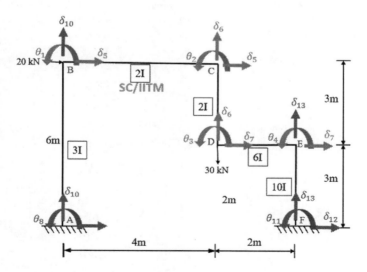

FIGURE 1.42 Unrestrained and restrained degrees-of-freedom.

Thus, the size of the unrestrained submatrix will be 7×7. The total size of the stiffness matrix will be 13×13. The element stiffness matrix will be 4×4, neglecting axial deformation.

The orientation and the labels of the members are given in the following table:

Member	jth end	kth end	Degrees-of-freedom labels
BA	B	A	1, 8, 5, 9
BC	B	C	1, 2, 10, 6
CD	C	D	2, 3, 5, 7
DE	D	E	3, 4, 6, 13
EF	E	F	4, 11, 7, 12

2. *Formulation of stiffness matrix:*

The stiffness matrix is given by,

$$K_i = \begin{bmatrix} \dfrac{4EI}{l} & \dfrac{2EI}{l} & \dfrac{6EI}{l^2} & -\dfrac{6EI}{l^2} \\ \dfrac{2EI}{l} & \dfrac{4EI}{l} & \dfrac{6EI}{l^2} & -\dfrac{6EI}{l^2} \\ \dfrac{6EI}{l^2} & \dfrac{6EI}{l^2} & \dfrac{12EI}{l^3} & -\dfrac{12EI}{l^3} \\ -\dfrac{6EI}{l^2} & -\dfrac{6EI}{l^2} & -\dfrac{12EI}{l^3} & \dfrac{12EI}{l^3} \end{bmatrix}$$

Thus, the size of the stiffness matrix is 4×4, neglecting axial deformation. For this problem, the element stiffness matrices are given as follows, considering the difference in the moment of inertia of all the members in the frame.

$$K_{AB} = EI \begin{bmatrix} 2 & 1 & 0.5 & -0.5 \\ 1 & 2 & 0.5 & -0.5 \\ 0.5 & 0.5 & 0.167 & -0.167 \\ -0.5 & -0.5 & -0.167 & 0.167 \end{bmatrix} \begin{matrix} 1 \\ 8 \\ 5 \\ 9 \end{matrix}$$

columns: ① ⑧ ⑤ ⑨

$$K_{BC} = EI \begin{bmatrix} 2 & 1 & 0.75 & -0.75 \\ 1 & 2 & 0.75 & -0.75 \\ 0.75 & 0.75 & 0.375 & -0.375 \\ -0.75 & -0.75 & -0.375 & 0.375 \end{bmatrix} \begin{matrix} 1 \\ 2 \\ 10 \\ 6 \end{matrix}$$

columns: ① ② ⑩ ⑥

Planar Orthogonal Structures

$$K_{CD} = EI \begin{bmatrix} 2.667 & 1.333 & 1.333 & -1.333 \\ 1.333 & 2.667 & 1.333 & -1.333 \\ 1.333 & 1.333 & 0.889 & -0.889 \\ -1.333 & -1.333 & -0.889 & 0.889 \end{bmatrix} \begin{matrix} ②\\③\\⑤\\⑦ \end{matrix}$$

(columns: ②, ③, ⑤, ⑦)

$$K_{DE} = EI \begin{bmatrix} 12 & 6 & 9 & -9 \\ 6 & 12 & 9 & -9 \\ 9 & 9 & 9 & -9 \\ -9 & -9 & -9 & 9 \end{bmatrix} \begin{matrix} ③\\④\\⑥\\⑬ \end{matrix}$$

(columns: ③, ④, ⑥, ⑬)

$$K_{EF} = EI \begin{bmatrix} 13.333 & 6.667 & 6.667 & -6.667 \\ 6.667 & 13.333 & 6.667 & -6.667 \\ 6.667 & 6.667 & 4.445 & -4.445 \\ -6.667 & -6.667 & -4.445 & 4.445 \end{bmatrix} \begin{matrix} ④\\⑪\\⑦\\⑫ \end{matrix}$$

(columns: ④, ⑪, ⑦, ⑫)

We need only the stiffness matrix of unrestrained degrees-of-freedom.

$$K_{UU} = EI \begin{bmatrix} 4 & 1 & 0 & 0 & 0.5 & -0.75 & 0 \\ 1 & 4.667 & 1.333 & 0 & 1.333 & -0.75 & -1.333 \\ 0 & 1.333 & 14.667 & 6 & 1.333 & 9 & -1.333 \\ 0 & 0 & 6 & 25.333 & 0 & 9 & 6.667 \\ 0.5 & 1.333 & 1.333 & 0 & 1.056 & 0 & -0.889 \\ -0.75 & -0.75 & 9 & 9 & 0 & 9.375 & 0 \\ 0 & -1.333 & -1.333 & 6.667 & -0.889 & 0 & 5.334 \end{bmatrix} \begin{matrix} ①\\②\\③\\④\\⑤\\⑥\\⑦ \end{matrix}$$

(columns: ①, ②, ③, ④, ⑤, ⑥, ⑦)

The size of the unrestrained stiffness matrix is 7×7. The inverse of the matrix is calculated using MATLAB programming and the solution is obtained by MATLAB code written for solving the planar orthogonal structures.

$$K_{UU}^{-1} = \frac{1}{EI} \begin{bmatrix} 0.2758 & -0.0298 & -0.0053 & -0.0016 & -0.1070 & 0.0263 & -0.0246 \\ -0.0298 & 0.3843 & -0.0747 & -0.0808 & -0.2639 & 0.1776 & 0.1344 \\ -0.0053 & -0.0747 & 0.2963 & 0.1098 & -0.4029 & -0.3962 & -0.1490 \\ -0.0016 & -0.0808 & 0.1098 & 0.1914 & -0.269 & -0.2957 & -0.2768 \\ -0.107 & -0.2639 & -0.4029 & -0.2690 & 2.3068 & 0.6154 & 0.5540 \\ 0.0263 & 0.1776 & -0.3962 & -0.2957 & 0.6154 & 0.7873 & 0.4175 \\ -0.0246 & 0.1344 & -0.1490 & -0.2768 & 0.5540 & 0.4175 & 0.6221 \end{bmatrix}$$

3. *Calculation of fixed end moments:*

Since the load is acting only on the joints, the joint load vector can be written directly.

4. *Calculation of joint load vectors:*

The size of the joint load vector is 13×1.

$$\{J_L\}_{13\times 1} = \begin{Bmatrix} 0 \\ 0 \\ 0 \\ 0 \\ 20 \\ -30 \\ 0 \\ \hline 0 \\ 0 \\ 0 \\ 0 \\ 0 \\ 0 \end{Bmatrix} \begin{matrix} 1 \\ 2 \\ 3 \\ 4 \\ 5 \\ 6 \\ 7 \\ 8 \\ 9 \\ 10 \\ 11 \\ 12 \\ 13 \end{matrix}$$

The joint load vector can be split into unrestrained and restrained degrees-of-freedom.

$$\{J_L\} = \begin{Bmatrix} (J_L)_u \\ _{7\times 1} \\ --- \\ (J_L)_r \\ _{6\times 1} \end{Bmatrix}_{13\times 1}$$

Thus,
$$J_{Lu} = \begin{Bmatrix} 0 \\ 0 \\ 0 \\ 0 \\ 20 \\ -30 \\ 0 \end{Bmatrix} \begin{matrix} 1 \\ 2 \\ 3 \\ 4 \\ 5 \\ 6 \\ 7 \end{matrix}$$

5. *Calculation of displacements:*

Now, $\{\Delta_u\} = [k_{uu}]^{-1} \{J_L\}_u$

Planar Orthogonal Structures

$$\begin{Bmatrix} \theta_1 \\ \theta_2 \\ \theta_3 \\ \theta_4 \\ \delta_5 \\ \delta_6 \\ \delta_7 \end{Bmatrix} = \frac{1}{EI} \begin{Bmatrix} -2.9291 \\ -10.6047 \\ 3.8284 \\ 3.4921 \\ 27.6762 \\ -11.3104 \\ -1.4465 \end{Bmatrix} \text{ radians}$$

6. **Calculation of end moment and shear:**

The general equation to find the end moment of the ith beam is,

$$[M]_i = k_i \delta_i + (\text{FEM})_i$$

The equation can be rewritten element-wise as follows:

$$M_{AB} = [k]_{AB} \{\delta\} + \{\text{FEM}\}$$

$$\begin{Bmatrix} M_1 \\ M_8 \\ V_5 \\ V_9 \end{Bmatrix} = EI \begin{bmatrix} 2 & 1 & 0.5 & -0.5 \\ 1 & 2 & 0.5 & -0.5 \\ 0.5 & 0.5 & 0.167 & -0.167 \\ -0.5 & -0.5 & -0.167 & 0.167 \end{bmatrix} \begin{Bmatrix} \theta_1 \\ \theta_8 \\ \delta_5 \\ \delta_9 \end{Bmatrix} + 0$$

Thus, $\begin{Bmatrix} M_1 \\ M_8 \\ V_5 \\ V_9 \end{Bmatrix} = \begin{Bmatrix} 7.98 \\ 10.9091 \\ 3.1482 \\ -3.1482 \end{Bmatrix}$

$$M_{BC} = [k]_{BC} \{\delta\} + \{\text{FEM}\}$$

$$\begin{Bmatrix} M_1 \\ M_2 \\ V_{10} \\ V_6 \end{Bmatrix} = EI \begin{bmatrix} 2 & 1 & 0.75 & -0.75 \\ 1 & 2 & 0.75 & -0.75 \\ 0.75 & 0.75 & 0.375 & -0.375 \\ -0.75 & -0.75 & -0.375 & 0.375 \end{bmatrix} \begin{Bmatrix} \theta_1 \\ \theta_2 \\ \delta_{10} \\ \delta_6 \end{Bmatrix} + 0 = \begin{Bmatrix} -7.98 \\ -15.6556 \\ -5.9089 \\ 5.9089 \end{Bmatrix}$$

Similarly, for the member CD, $M_{CD} = [k]_{CD} \{\delta\} + \{\text{FEM}\}$

$$\begin{Bmatrix} M_2 \\ M_3 \\ V_5 \\ V_7 \end{Bmatrix} = EI \begin{bmatrix} 2.667 & 1.333 & 1.333 & -1.333 \\ 1.333 & 2.667 & 1.333 & -1.333 \\ 1.333 & 1.333 & 0.889 & -0.889 \\ -1.333 & -1.333 & -0.889 & 0.889 \end{bmatrix} \begin{Bmatrix} \theta_2 \\ \theta_3 \\ \delta_5 \\ \delta_7 \end{Bmatrix} + 0 = \begin{Bmatrix} 15.6556 \\ 34.8998 \\ 16.8518 \\ -16.8518 \end{Bmatrix}$$

$$M_{DE} = [k]_{DE}\{\delta\} + \{FEM\}$$

$$\begin{Bmatrix} M_3 \\ M_4 \\ V_6 \\ V_{13} \end{Bmatrix} = EI \begin{bmatrix} 12 & 6 & 9 & -9 \\ 6 & 12 & 9 & -9 \\ 9 & 9 & 9 & -9 \\ -9 & -9 & -9 & 9 \end{bmatrix} \begin{Bmatrix} \theta_3 \\ \theta_4 \\ \delta_6 \\ \delta_{13} \end{Bmatrix} + 0 = \begin{Bmatrix} -34.8998 \\ -36.9180 \\ -35.9089 \\ 35.9089 \end{Bmatrix}$$

$$M_{EF} = [k]_{EF}\{\delta\} + \{FEM\}$$

$$\begin{Bmatrix} M_4 \\ M_{11} \\ V_7 \\ V_{12} \end{Bmatrix} = EI \begin{bmatrix} 13.333 & 6.667 & 6.667 & -6.667 \\ 6.667 & 13.333 & 6.667 & -6.667 \\ 6.667 & 6.667 & 4.445 & -4.445 \\ -6.667 & -6.667 & -4.445 & 4.445 \end{bmatrix} \begin{Bmatrix} \theta_4 \\ \theta_{11} \\ \delta_7 \\ \delta_{12} \end{Bmatrix} + 0 = \begin{Bmatrix} 36.9180 \\ 13.6375 \\ 16.8518 \\ -16.8518 \end{Bmatrix}$$

The member and final end moments and shear are shown in Figures 1.43 and 1.44.

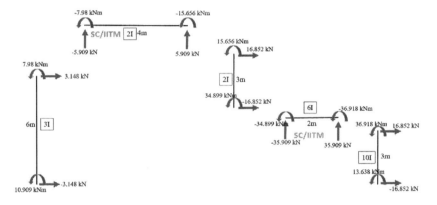

FIGURE 1.43 Member end moments and shear.

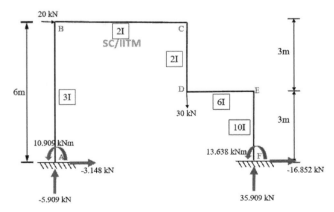

FIGURE 1.44 Final end moments and shear.

Planar Orthogonal Structures

1.8.6 Computer Program for Step Frame

```
%% stiffness matrix method
% Input
clc;
clear;
n = 5; % number of members
I = [3 2 2 6 10]; %Moment of inertis in m4
L = [6 4 3 2 3]; % length in m
uu = 7; % Number of unrestrained degrees of freedom
ur = 6; % Number of restrained degrees of freedom
uul = [1 2 3 4 5 6 7]; % global labels of unrestrained dof
url = [8 9 10 11 12 13]; % global labels of restrained dof
l1 = [1 8 5 9]; % Global labels for member 1
l2 = [1 2 10 6]; % Global labels for member 2
l3 = [2 3 5 7]; % Global labels for member 3
l4 = [3 4 6 13]; % Global labels for member 4
l5 = [4 11 7 12]; % Global labels for member 5
l= [l1; l2; l3; l4; l5];
dof = uu+ur;
Ktotal = zeros (dof);
fem1= zeros (1,4); % Local Fixed end moments of member 1
fem2= zeros (1,4); % Local Fixed end moments of member 2
fem3= zeros (1,4); % Local Fixed end moments of member 3
fem4= zeros (1,4); % Local Fixed end moments of member 4
fem5= zeros (1,4); % Local Fixed end moments of member 5

%% rotation coefficients for each member
rc1 = 4.*I./L;
rc2 = 2.*I./L;

%% stiffness matrix 4 by 4 (axial deformation neglected)
for i = 1:n
     Knew = zeros (dof);
     k1 = [rc1(i); rc2(i); (rc1(i)+rc2(i))/L(i); 
(-(rc1(i)+rc2(i))/L(i))];
     k2 = [rc2(i); rc1(i); (rc1(i)+rc2(i))/L(i); 
(-(rc1(i)+rc2(i))/L(i))];
     k3 = [(rc1(i)+rc2(i))/L(i); (rc1(i)+rc2(i))/L(i); 
(2*(rc1(i)+rc2(i))/(L(i)^2)); (-2*(rc1(i)+rc2(i))/(L(i)^2))];
     k4 = -k3;
     K = [k1 k2 k3 k4];
     fprintf ('Member Number =');
     disp (i);
     fprintf ('Local Stiffness matrix of member, [K] = \n');
     disp (K);
     for p = 1:4
          for q = 1:4
              Knew((l(i,p)),(l(i,q))) =K(p,q);
          end
```

```
            end
            Ktotal = Ktotal + Knew;
            if i == 1
                Kg1=K;
            elseif i == 2
                Kg2 = K;
            elseif i ==3
                Kg3=K;
            elseif i== 4
                Kg4 = K;
            else
                Kg5 = K;
            end
end
fprintf ('Stiffness Matrix of complete structure, [Ktotal] = \n');
disp (Ktotal);
Kunr = zeros(uu);
for x=1:uu
    for y=1:uu
        Kunr(x,y)= Ktotal(x,y);
    end
end
fprintf ('Unrestrained Stiffness sub-matrix, [Kuu] = \n');
disp (Kunr);
KuuInv= inv(Kunr);
fprintf ('Inverse of Unrestrained Stiffness sub-matrix,
[KuuInverse] = \n');
disp (KuuInv);

%% Creation of joint load vector
jl= [0; 0; 0; 0; 20; -30; 0; 0; 0; 0; 0; 0; 0]; % values given
in kN or kNm
jlu = [0; 0; 0; 0; 20; -30; 0]; % load vector in unrestrained dof
delu = KuuInv*jlu;
fprintf ('Joint Load vector, [Jl] = \n');
disp (jl');
fprintf ('Unrestrained displacements, [DelU] = \n');
disp (delu');
delr = zeros (ur,1);
del = [delu; delr];
deli= zeros (4,1);
for i = 1:n
    for p = 1:4
        deli(p,1) = del((l(i,p)),1) ;
    end
    if i == 1
        delbar1 = deli;
        mbar1= (Kg1 * delbar1)+fem1';
        fprintf ('Member Number =');
        disp (i);
        fprintf ('Global displacement matrix [DeltaBar] = \n');
```

Planar Orthogonal Structures

```
            disp (delbar1');
            fprintf ('Global End moment matrix [MBar] = \n');
            disp (mbar1');
        elseif i == 2
            delbar2 = deli;
            mbar2= (Kg2 * delbar2)+fem2';
            fprintf ('Member Number =');
            disp (i);
            fprintf ('Global displacement matrix [DeltaBar] = \n');
            disp (delbar2');
            fprintf ('Global End moment matrix [MBar] = \n');
            disp (mbar2');
        elseif i == 3
            delbar3 = deli;
            mbar3= (Kg3 * delbar3)+fem3';
            fprintf ('Member Number =');
            disp (i);
            fprintf ('Global displacement matrix [DeltaBar] = \n');
            disp (delbar3');
            fprintf ('Global End moment matrix [MBar] = \n');
            disp (mbar3');
        elseif i == 4
            delbar4 = deli;
            mbar4= (Kg4 * delbar4)+fem4';
            fprintf ('Member Number =');
            disp (i);
            fprintf ('Global displacement matrix [DeltaBar] = \n');
            disp (delbar4');
            fprintf ('Global End moment matrix [MBar] = \n');
            disp (mbar4');
        else
            delbar5 = deli;
            mbar5= (Kg5 * delbar5)+fem5';
            fprintf ('Member Number =');
            disp (i);
            fprintf ('Global displacement matrix [DeltaBar] = \n');
            disp (delbar5');
            fprintf ('Global End moment matrix [MBar] = \n');
            disp (mbar5');
        end
end
%% check
mbar = [mbar1'; mbar2'; mbar3'; mbar4'; mbar5'];
jf = zeros(dof,1);
for a=1:n
    for b=1:4 % size of k matrix
        d = l(a,b);
        jfnew = zeros(dof,1);
        jfnew(d,1)=mbar(a,b);
        jf=jf+jfnew;
    end
```

```
end
fprintf ('Joint forces = \n');
disp (jf');
```

MATLAB output:

```
Member Number =  1
Local Stiffness matrix of member, [K] =

    2.0000    1.0000    0.5000   -0.5000
    1.0000    2.0000    0.5000   -0.5000
    0.5000    0.5000    0.1667   -0.1667
   -0.5000   -0.5000   -0.1667    0.1667

Member Number =  2
Local Stiffness matrix of member, [K] =

    2.0000    1.0000    0.7500   -0.7500
    1.0000    2.0000    0.7500   -0.7500
    0.7500    0.7500    0.3750   -0.3750
   -0.7500   -0.7500   -0.3750    0.3750

Member Number =  3
Local Stiffness matrix of member, [K] =

    2.6667    1.3333    1.3333   -1.3333
    1.3333    2.6667    1.3333   -1.3333
    1.3333    1.3333    0.8889   -0.8889
   -1.3333   -1.3333   -0.8889    0.8889

Member Number =  4
Local Stiffness matrix of member, [K] =

      12        6        9       -9
       6       12        9       -9
       9        9        9       -9
      -9       -9       -9        9

Member Number =  5
Local Stiffness matrix of member, [K] =

   13.3333    6.6667    6.6667   -6.6667
    6.6667   13.3333    6.6667   -6.6667
    6.6667    6.6667    4.4444   -4.4444
   -6.6667   -6.6667   -4.4444    4.4444

Stiffness Matrix of complete structure, [Ktotal] =

   4.0000  1.0000        0        0   0.5000  -0.7500        0   1.0000  -0.5000   0.7500        0
   1.0000  4.6667   1.3333        0   1.3333  -0.7500  -1.3333        0        0   0.7500        0
        0  1.3333  14.6667   6.0000   1.3333   9.0000  -1.3333        0        0        0        0
        0       0   6.0000  25.3333        0   9.0000   6.6667        0        0        0   6.6667
   0.5000  1.3333   1.3333        0   1.0556        0  -0.8889   0.5000  -0.1667        0        0
```

Planar Orthogonal Structures

```
-0.7500 -0.7500  9.0000  9.0000       0  9.3750       0       0       0 -0.3750       0
      0 -1.3333 -1.3333  6.6667 -0.8889       0  5.3333       0       0       0  6.6667
 1.0000       0       0       0  0.5000       0       0  2.0000 -0.5000       0       0
-0.5000       0       0       0 -0.1667       0       0 -0.5000  0.1667       0       0
 0.7500  0.7500       0       0       0 -0.3750       0       0       0  0.3750       0
      0       0       0  6.6667       0       0  6.6667       0       0       0 13.3333
      0       0       0 -6.6667       0       0 -4.4444       0       0       0 -6.6667
      0       0 -9.0000 -9.0000       0 -9.0000       0       0       0       0       0
```

Columns 12 through 13

```
       0        0
       0        0
       0  -9.0000
 -6.6667  -9.0000
       0        0
       0  -9.0000
 -4.4444        0
       0        0
       0        0
       0        0
 -6.6667        0
  4.4444        0
       0   9.0000
```

Unrestrained Stiffness sub-matrix, [Kuu] =

```
 4.0000  1.0000       0       0  0.5000 -0.7500       0
 1.0000  4.6667  1.3333       0  1.3333 -0.7500 -1.3333
      0  1.3333 14.6667  6.0000  1.3333  9.0000 -1.3333
      0       0  6.0000 25.3333       0  9.0000  6.6667
 0.5000  1.3333  1.3333       0  1.0556       0 -0.8889
-0.7500 -0.7500  9.0000  9.0000       0  9.3750       0
      0 -1.3333 -1.3333  6.6667 -0.8889       0  5.3333
```

Inverse of Unrestrained Stiffness sub-matrix, [KuuInverse] =

```
 0.2758 -0.0298 -0.0053 -0.0016 -0.1070  0.0263 -0.0246
-0.0298  0.3843 -0.0747 -0.0808 -0.2639  0.1776  0.1344
-0.0053 -0.0747  0.2963  0.1098 -0.4029 -0.3962 -0.1490
-0.0016 -0.0808  0.1098  0.1914 -0.2690 -0.2957 -0.2768
-0.1070 -0.2639 -0.4029 -0.2690  2.3068  0.6154  0.5540
 0.0263  0.1776 -0.3962 -0.2957  0.6154  0.7873  0.4175
-0.0246  0.1344 -0.1490 -0.2768  0.5540  0.4175  0.6221
```

Joint Load vector, [Jl] =

```
  0   0   0   0  20 -30   0   0   0   0   0   0
```

Unrestrained displacements, [DelU] =

```
 -2.9291 -10.6047  3.8284  3.4921 27.6762 -11.3104 -1.4465
```

```
Member Number = 1
Global displacement matrix [DeltaBar] =
   -2.9291      0     27.6762      0
Global End moment matrix [MBar] =
    7.9800    10.9091     3.1482    -3.1482
Member Number = 2
Global displacement matrix [DeltaBar] =
   -2.9291   -10.6047       0    -11.3104
Global End moment matrix [MBar] =
   -7.9800   -15.6556    -5.9089     5.9089
Member Number = 3
Global displacement matrix [DeltaBar] =
  -10.6047     3.8284    27.6762    -1.4465
Global End moment matrix [MBar] =
   15.6556    34.8998    16.8518   -16.8518
Member Number = 4
Global displacement matrix [DeltaBar] =
    3.8284     3.4921   -11.3104       0
Global End moment matrix [MBar] =
  -34.8998   -36.9180   -35.9089    35.9089
Member Number = 5
Global displacement matrix [DeltaBar] =
    3.4921      0     -1.4465      0
Global End moment matrix [MBar] =
   36.9180    13.6375    16.8518   -16.8518
Joint forces =
     0.0000   -0.0000   -0.0000   -0.0000   20.0000  -30.0000   -0.0000
    10.9091   -3.1482   -5.9089   13.6375  -16.8518   35.9089
```

2 Planar Non-Orthogonal Structures

2.1 PLANAR NON-ORTHOGONAL STRUCTURE

The analysis of a simple single story single bay frame is much easier through the stiffness method. The method becomes more complicated, when the members in the structure are not orthogonal to each other. A typical example of such a structure is a jacket platform, which is a template structure acted upon by wave forces. The members in a jacket platform are always not orthogonal to each other.

Consider the conventional equation for a complete structure,

$$[K]_c \{\Delta\}_c = \{J_L\}_c + [R]_c \tag{2.1}$$

The previous equation is valid only when the local axes system of the member and reference axes system of the complete structure matches. In such a condition, the stiffness matrix can be established in terms of submatrices by partitioning them and solving them for displacements and moments. When the reference axes of the system and the member axes of the system are not aligned, then the stiffness matrix of each member may not be the same for the members not aligned with the reference axes of the structural system. Thus, the local stiffness matrix needs to be transformed with reference to global axes system to solve the problem. This is the difficulty in solving non-orthogonal structural systems. Thus, the complication starts only when the members are non-orthogonal to each other. Hence, all members must be aligned or transformed with respect to the reference axes of the system. The individual stiffness matrices of each member should be developed in the same manner as discussed earlier. But, it should be transformed within the frame of reference axes of the system. The reference axes or global axes system and the local axis system are represented as shown in Figure 2.1.

There has got to be a connectivity established between the local axis and the global axis to solve a problem. The derivations discussed before, for the members aligned to the local axis system of the member, is completely valid. But, they need to be transformed when the local axis system is not mapped exactly to the reference axis system. If the local and reference axes system matches, there is no need of transformation. But, it is possible only when the members are orthogonal to each other. When the members are non-orthogonal to each other, then the mapping of the local axes system to the reference axes system is a major issue. The transformation has to be done on both the ways between the local axes and the global axes, which is required for the design of the members. Thus, the procedure has to be established for the development of the transformation matrix both ways.

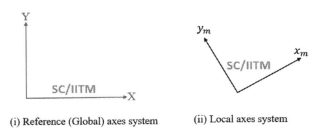

FIGURE 2.1 Global and local axis system.

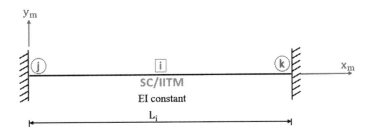

FIGURE 2.2 Local axes system representation for a member.

The stiffness matrices developed so far, are with reference to the local axes system. Consider a member as shown in Figure 2.2, with nodes j and k. The local axes system of the member is represented by the x_m, y_m axes. One must choose the x_m axis such that the length of the member lies on the positive side. Thus, it all depends on the location of the jth end of the member. Subsequently, the kth end of the member is located on the positive side of the x_m axis. The y_m axis is located anticlockwise or counterclockwise by 90° to x_m. The length of the member is located on the positive side of x_m.

The stiffness matrix derivation done so far, is valid for the local axes system. It is very important to note that the choice of the jth end helps to orient the member local axes to that of the reference axes.

For example, consider an orthogonal system and non-orthogonal systems, as shown in Figure 2.3. The reference axes for both the systems are marked the same. But, the local axes system will vary for both members. All conditions mentioned previously while choosing the jth end and local axes should be satisfied. Thus, orienting the member with reference to the global axis will happen by choosing the jth and kth end of each member. When the members are non-orthogonal, it can be seen that the local axes of the members are not aligned with the reference axes of the structural system. Thus, the members need to be transformed to the reference axes of the structural system.

In case of non-orthogonal frames, orientation of the local axes (x_m-y_m) may be such that this cannot be aligned or mapped to the same orientation of the reference axes (X-Y) (or example, members 1 and 3 in Figure 2.3). Hence, the member stiffness matrix cannot be directly written with respect to the reference axes. But, there

Planar Non-Orthogonal Structures

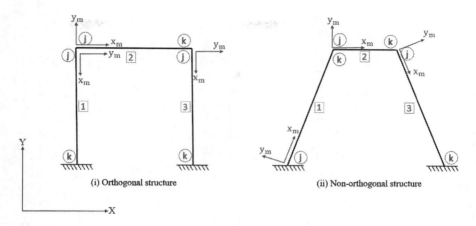

FIGURE 2.3   Global and local axes of the structural systems.

is a solution for this problem. It can be transformed to the reference axes system. The stiffness matrix written for the local axes system will be valid. We need to transform the same for the reference axes system. This is true when the members are non-orthogonal. No additional efforts are required to derive or compute the stiffness matrix in $(x_m\text{-}y_m)$ frame. The methods used earlier are all valid and only transformation has to be done. Thus, $[K_i]_{x_m-y_m}$ needs a transformation into the reference axes $(X\text{-}Y)$. $[K]_{X-Y}$ cannot be written directly for non-orthogonal members.

2.2   STIFFNESS MATRIX FORMULATION

Let us now consider a beam element, which is slightly modified to be a member of non-orthogonal frames, as shown in Figure 2.4. Here, the axial deformation is also included. The displacements at both the ends are marked along with the local axes system. There are six degrees-of-freedom. Now, the stiffness matrix will be of size 6×6.

We already know the submatrix in the stiffness matrix neglecting the axial deformation. It is given by,

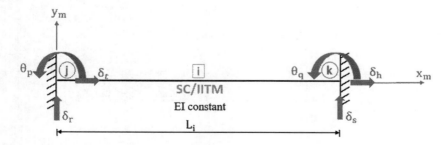

FIGURE 2.4   Beam element.

$$[k] = \begin{bmatrix} k_{pp} & k_{pq} & \dfrac{k_{pp}+k_{pq}}{L} & -\left(\dfrac{k_{pp}+k_{pq}}{L}\right) \\ k_{qp} & k_{qq} & \dfrac{k_{qp}+k_{qq}}{L} & -\left(\dfrac{k_{qp}+k_{qq}}{L}\right) \\ \dfrac{k_{pp}+k_{pq}}{L} & \dfrac{k_{pq}+k_{qq}}{L} & \dfrac{k_{pp}+k_{pq}+k_{qp}+k_{qq}}{L^2} & \dfrac{k_{pp}+k_{pq}+k_{qp}+k_{qq}}{L^2} \\ -\left(\dfrac{k_{pp}+k_{pq}}{L}\right) & -\left(\dfrac{k_{pq}+k_{qq}}{L}\right) & -\left(\dfrac{k_{pp}+k_{pq}+k_{qp}+k_{qq}}{L^2}\right) & -\left(\dfrac{k_{pp}+k_{pq}+k_{qp}+k_{qq}}{L^2}\right) \end{bmatrix}$$

To develop the stiffness matrix for the non-orthogonal frame, consider the beam element undergoing unit axial deformation, as shown in Figure 2.5.

k_{tt} is the force in the tth degree by giving unit displacement in the tth degree.
k_{ht} is the force in the hth degree by giving unit displacement in the tth degree.

Here, $k_{tt} = \dfrac{AE}{l}\delta_u = \dfrac{AE}{l}$ (Since, the beam is assumed to undergo unit displacement)

$$k_{ht} = -\dfrac{AE}{l} \tag{2.2}$$

Similarly, assuming the beam element undergoing axial deformation on the other end, as shown in Figure 2.6.

$$k_{th} = -\dfrac{AE}{l} \tag{2.3}$$

$$k_{hh} = \dfrac{AE}{l}\delta_u = \dfrac{AE}{l} \tag{2.4}$$

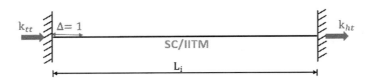

FIGURE 2.5 Beam element with axial deformation.

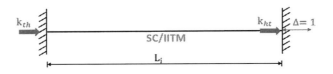

FIGURE 2.6 Beam element with axial deformation.

Planar Non-Orthogonal Structures

Thus, the matrix involving axial deformation can be written as follows:

$$\begin{bmatrix} \dfrac{AE}{l} & -\dfrac{AE}{l} \\ -\dfrac{AE}{l} & \dfrac{AE}{l} \end{bmatrix}$$

By combining the previously mentioned stiffness submatrix for axial deformation with the stiffness matrix derived before, one can get the full stiffness matrix of the non-orthogonal structural system.

Thus, the complete stiffness matrix for the member at the local axes is given by,

$$[K]_i = \begin{bmatrix} \dfrac{4EI}{l} & \dfrac{2EI}{l} & \dfrac{6EI}{l^2} & -\dfrac{6EI}{l^2} & 0 & 0 \\ \dfrac{2EI}{l} & \dfrac{4EI}{l} & \dfrac{6EI}{l^2} & -\dfrac{6EI}{l^2} & 0 & 0 \\ \dfrac{6EI}{l^2} & \dfrac{6EI}{l^2} & \dfrac{12EI}{l^3} & -\dfrac{12EI}{l^3} & 0 & 0 \\ -\dfrac{6EI}{l^2} & -\dfrac{6EI}{l^2} & -\dfrac{12EI}{l^3} & \dfrac{12EI}{l^3} & 0 & 0 \\ 0 & 0 & 0 & 0 & \dfrac{AE}{l} & -\dfrac{AE}{l} \\ 0 & 0 & 0 & 0 & -\dfrac{AE}{l} & \dfrac{AE}{l} \end{bmatrix}$$

This is the complete 6×6 stiffness matrix of the member at the local axes. If the local axes of the member does not orient with the global axes or the reference axes of the system, transformation has to be done. Thus, the transformation matrix has to be derived to transform the previously mentioned stiffness matrix to the reference axes system.

2.3 TRANSFORMATION MATRIX

Let us consider two orthogonal sets of axes, such as $(x_1\text{-}y_1)$ and $(x_2\text{-}y_2)$, as shown in Figure 2.7.

It can be seen that y_1 is anticlockwise 90 degrees to x_1. Similarly, y_2 is anticlockwise 90 degrees to x_2. Both the axes have a common origin. Let $(x_2\text{-}y_2)$ be rotated anticlockwise by θ degrees. The horizontal and vertical components in $(x_1\text{-}y_1)$ axes are $\overline{V_1}$ and $\overline{V_2}$ respectively. One can resolve $(x_1\text{-}y_1)$ to $(x_2\text{-}y_2)$ axes. Thus,

$$V_1 = \overline{V_1}\cos\theta + \overline{V_2}\sin\theta$$

$$V_2 = \overline{V_2}\cos\theta - \overline{V_1}\sin\theta \tag{2.5}$$

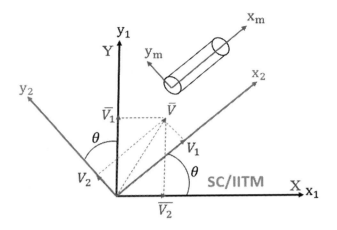

FIGURE 2.7 Two orthogonal sets of axes.

The previous equations can be written in matrix form as follows:

$$\begin{Bmatrix} V_1 \\ V_2 \end{Bmatrix} = \begin{bmatrix} \cos\theta & \sin\theta \\ -\sin\theta & \cos\theta \end{bmatrix} \begin{Bmatrix} \overline{V}_1 \\ \overline{V}_2 \end{Bmatrix}$$

It can also be expressed as follows:

$$\{V\} = [T]\{\overline{V}\} \tag{2.6}$$

Where, $[T]$ is the transformation matrix.

Alternatively, one can also resolve $(x_2\text{-}y_2)$ to $(x_1\text{-}y_1)$ axes. In this case,

$$\overline{V}_1 = V_1 \cos\theta - V_2 \sin\theta$$

$$\overline{V}_2 = V_2 \cos\theta + V_1 \sin\theta \tag{2.7}$$

Expressing the previous equations in matrix form,

$$\begin{Bmatrix} \overline{V}_1 \\ \overline{V}_2 \end{Bmatrix} = \begin{bmatrix} \cos\theta & -\sin\theta \\ \sin\theta & \cos\theta \end{bmatrix} \begin{Bmatrix} V_1 \\ V_2 \end{Bmatrix}$$

$$\{\overline{V}\} = [T]^T \{V\} \tag{2.8}$$

Planar Non-Orthogonal Structures

The transformation between the axes in both ways is valid. Thus, the transformation matrix is given by,

$$[T] = \begin{bmatrix} c & s \\ -s & c \end{bmatrix}$$

Where,
   c represents $\cos(\theta)$
   s represents $\sin(\theta)$

Here, θ is the angle between the two axes measured in a specific style. For any arbitrarily oriented member, local axes (x_m-y_m) should be parallel to (x_2-y_2), for which reference axes will be (x_1-y_1). Hence, θ is the inclination or rotation of x_m with respect to x measured in an anticlockwise manner. We already know how to mark the local axes for a given member that is arbitrarily oriented with respect to the reference axes. X_m should be considered in such a manner, so that the length of the member should be on the positive side of x_m and y_m should be 90 degrees anticlockwise to x_m.

Important property of transformation matrix:
The inverse of the transformation matrix is given by,

$$[T]^{-1} = \frac{1}{\cos^2\theta + \sin^2\theta} \begin{bmatrix} \cos\theta & -\sin\theta \\ \sin\theta & \cos\theta \end{bmatrix} = [T]^T$$

Thus, the inverse of the transformation matrix is equal to the transpose of the matrix. Hence, the transformation matrix is orthogonal.

2.4 TRANSFORMATION MATRIX FOR END MOMENTS

Let us now consider a beam that is arbitrarily oriented with respect to the reference axes, as shown in Figure 2.8. The member is inclined by an angle θ. The member also undergoes axial deformation and hence the degree-of-freedom is six. Now, let

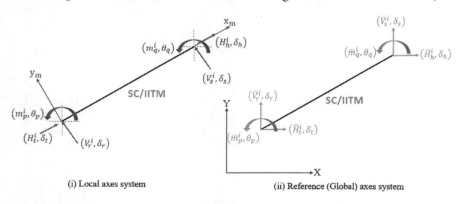

FIGURE 2.8 Transformation matrix for end moments and displacements.

us map all the degrees-of-freedom with respect to the reference axes. The mapping of the displacements with respect to the reference axes will become parallel to the reference axes themselves.

Thus, displacements along both local and reference (global) axes are given by,

$$\text{Displacements along the reference axes} = \begin{Bmatrix} \bar{m}_p \\ \bar{m}_q \\ \bar{V}_r \\ \bar{V}_s \\ \bar{H}_t \\ \bar{H}_h \end{Bmatrix}$$

$$\text{Displacement along the local axes} = \begin{Bmatrix} m_p \\ m_q \\ V_r \\ V_s \\ H_t \\ H_h \end{Bmatrix}$$

The displacements on the local and reference axes are connected by the transformation matrix as follows:

$$\begin{Bmatrix} \bar{m}_p \\ \bar{m}_q \\ \bar{V}_r \\ \bar{V}_s \\ \bar{H}_t \\ \bar{H}_h \end{Bmatrix} = \begin{bmatrix} 1 & 0 & 0 & 0 & 0 & 0 \\ 0 & 1 & 0 & 0 & 0 & 0 \\ 0 & 0 & \cos\theta & 0 & \sin\theta & 0 \\ 0 & 0 & 0 & \cos\theta & 0 & \sin\theta \\ 0 & 0 & -\sin\theta & 0 & \cos\theta & 0 \\ 0 & 0 & 0 & -\sin\theta & 0 & \cos\theta \end{bmatrix} \begin{Bmatrix} m_p \\ m_q \\ V_r \\ V_s \\ H_t \\ H_h \end{Bmatrix}$$

Since,

$$\bar{m}_p = m_p$$

$$\bar{m}_q = m_q$$

$$\bar{V}_r = V_r \cos\theta + H_t \sin\theta$$

$$\bar{V}_r = V_s \cos\theta + H_h \sin\theta \qquad (2.9)$$

$$\bar{H}_t = -V_r \sin\theta + H_t \cos\theta$$

$$\bar{H}_h = -V_s \sin\theta + H_h \cos\theta$$

Thus, it can be said that $\bar{m}_i = T^T m_i$.

Thus, the transformation matrix is written as follows:

Planar Non-Orthogonal Structures

$$m_i = T\bar{m}_i \tag{2.10}$$

$$\begin{Bmatrix} m_p \\ m_q \\ V_r \\ V_s \\ H_t \\ H_h \end{Bmatrix} = \begin{bmatrix} 1 & 0 & 0 & 0 & 0 & 0 \\ 0 & 1 & 0 & 0 & 0 & 0 \\ 0 & 0 & \cos\theta & 0 & -\sin\theta & 0 \\ 0 & 0 & 0 & \cos\theta & 0 & -\sin\theta \\ 0 & 0 & \sin\theta & 0 & \cos\theta & 0 \\ 0 & 0 & 0 & \sin\theta & 0 & \cos\theta \end{bmatrix} \begin{Bmatrix} \bar{m}_p \\ \bar{m}_q \\ \bar{V}_r \\ \bar{V}_s \\ \bar{H}_t \\ \bar{H}_h \end{Bmatrix}$$

Thus, the following set of equations can be written as follows:

$$\{\bar{\delta}\}_i = [T]_i^T \{\delta\}_i$$

$$\{\delta\}_i = [T]_i \{\bar{\delta}_i\} \tag{2.11}$$

where,

$$\{\delta\}_i = \begin{Bmatrix} \theta_p \\ \theta_q \\ \delta_r \\ \delta_s \\ \delta_t \\ \delta_h \end{Bmatrix}, \quad \{\bar{\delta}\}_i = \begin{Bmatrix} \bar{\theta}_p \\ \bar{\theta}_q \\ \bar{\delta}_r \\ \bar{\delta}_s \\ \bar{\delta}_t \\ \bar{\delta}_h \end{Bmatrix}$$

It can be seen that the transformation matrix has sine and cosine values. The component resolved along the x axis is $\cos\theta$ and the component along the y axis is $\sin\theta$. They are denoted as C_x and C_y respectively. Thus, the transformation matrix can be finally written as follows:

$$[T] = \begin{bmatrix} 1 & 0 & 0 & 0 & 0 & 0 \\ 0 & 1 & 0 & 0 & 0 & 0 \\ 0 & 0 & C_x & 0 & -C_y & 0 \\ 0 & 0 & 0 & C_x & 0 & -C_y \\ 0 & 0 & C_y & 0 & C_x & 0 \\ 0 & 0 & 0 & C_y & 0 & C_x \end{bmatrix}$$

It is interesting to note that the local axes (x_m-y_m) are rotated anticlockwise by θ degrees with respect to the reference axes. Thus, θ is measured anticlockwise, which means that the angle is measured anticlockwise between positive x_m and X axes. Now, this will govern the orientation of the member with respect to the reference axes. The transformation matrix will automatically take care of the mapping.

2.5 GLOBAL STIFFNESS MATRIX

Let us consider an arbitrarily oriented member with respect to the reference axes, as shown in Figure 2.9: The length of the member L_i. The length of the member can be easily mapped with respect to the reference axes.

$$C_x = \cos\theta = \frac{(X_k - X_j)}{L_i}$$

$$C_y = \sin\theta = \frac{(Y_k - Y_j)}{L_i} \qquad (2.12)$$

$$L_i = \sqrt{(X_k - X_j)^2 + (Y_k - Y_j)^2}$$

Thus, knowing the values of C_x, C_y and θ, one can define the transformation matrix of the arbitrarily oriented member with reference to the reference axes. It can be said that, for the known orientation of $(x_m\text{-}y_m)$ axes with respect to the $(X\text{-}Y)$ axes, the transformation matrix is completely known. Since the stiffness matrix of the member in the local axes is already known, the stiffness matrix with respect to the global axes can be found.

We know,

$$\{m\}_i = [T]_i \{\bar{m}_i\} \qquad (2.13)$$

$$\{\delta\}_i = [T]_i \{\bar{\delta}_i\} \qquad (2.14)$$

Substituting the previous equations in the following equation,

$$\{m\}_i = [K]_i \{\delta_i\}$$

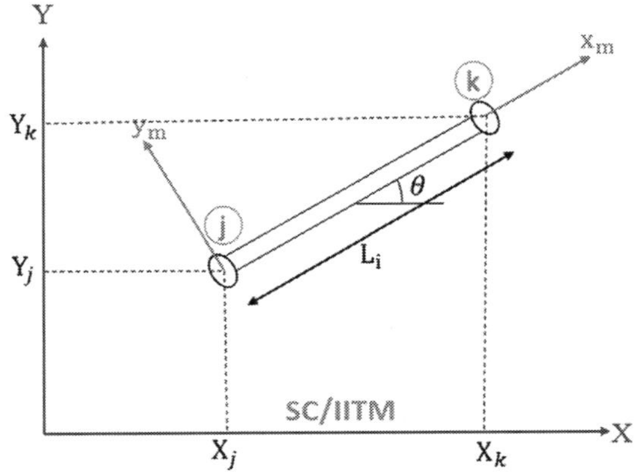

FIGURE 2.9 Transformation of length of the member.

Planar Non-Orthogonal Structures

$$[T]_i\{\bar{m}_i\} = [K]_i[T]_i\{\bar{\delta}_i\} \qquad (2.15)$$

Pre-multiplying with T^{-1},

$$\{\bar{m}_i\} = [T]_i^{-1}[K]_i[T]_i\{\bar{\delta}_i\} \qquad (2.16)$$

Since, $[T]$ is an orthogonal matrix, the previous equation can be rewritten as follows:

$$\{\bar{m}_i\} = [T]_i^T[K]_i[T]_i\{\bar{\delta}_i\} \qquad (2.17)$$

Thus, from the previous equation, the stiffness matrix of the member with respect to the reference axes can be written as,

$$\{\bar{K}\}_i = [T]_i^T[K]_i[T]_i \qquad (2.18)$$

The previously mentioned relationship can be used to find the global stiffness matrix of the ith member that is arbitrarily oriented with respect to the reference axes. Now, the following equation can be written, which gives the relationship between the component end displacements and end actions of the member in $(X-Y)$ axes system:

$$\{\bar{m}\}_i = [\bar{K}]_i\{\bar{\delta}_i\} \qquad (2.19)$$

Further,

$$[K]\{\Delta\} = \{J_L\} + \{R\}$$

$$[K_{uu}]\{\Delta_u\} = \{J_L\}_u \qquad (2.20)$$

$$[K_{ru}]\{\Delta_u\} - \{J_L\}_u = \{R\}_r$$

2.6 IMPORTANT STEPS IN ANALYSIS OF NON-ORTHOGONAL STRUCTURES

Step 1: Locate the local axes of the members
Choosing the jth end of the member will position the $(x_m\text{-}y_m)$ axes of the member. The origin of the axes $(x_m\text{-}y_m)$ is at the jth end. The x_m axis should be oriented toward the kth end making length of the member in the positive side of $x_m\text{-}y_m$ axis is 90 degrees anticlockwise to x_m. Now, this will fix the kth end.

Step 2: Locate reference axes and calculate θ
Locate the position of the reference axes $(X-Y)$. Find θ, which is the anticlockwise angle between $(X-Y)$ and the $(x_m\text{-}y_m)$ axes.

Step 3: Compute transformation matrix coefficients of each member
Since θ is known, find C_x and C_y and determine the transformation matrix $[T]$ for each member.

Step 4: Identify/label the degrees-of-freedom
Choose unrestrained displacements, both translational and rotational as the first group. Then choose the restrained displacement as the second group. The important point to be considered here is that the unrestrained and the restrained degrees-of-freedom are chosen with respect to the reference axes and not with respect to the local axes.

For example, consider a non-orthogonal frame, as shown in Figure 2.10. The reference axes is also marked for the frame. The degrees-of-freedom are marked with respect to the reference axes. There are nine unrestrained and six unrestrained degrees-of-freedom. The unrestrained degrees should be grouped first, followed by the restrained degrees. It can also be seen that the displacements, in all restrained and unrestrained degrees-of-freedom, are all oriented with respect to the reference axes (*X-Y*), but not with respect to the local axes.

Step 5: Estimate the stiffness matrix
The stiffness matrix, including the axial deformation, will be of size 6×6. Then, find the global stiffness matrix for every member. Finally, assemble the stiffness matrix for the complete structure and get the submatrix of unrestrained degrees-of-freedom from the total stiffness matrix.

Step 6: Calculation of fixed end moments and joint loads
Joint loads should be computed with respect to the reference axes only. Joint loads are the reversal of the sign of fixed end moments (FEM). Thus,

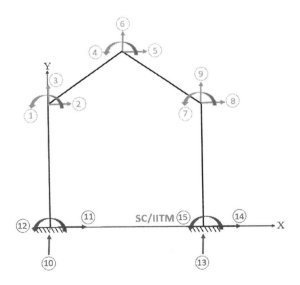

FIGURE 2.10 Marking degrees-of-freedom.

Planar Non-Orthogonal Structures

$$[\overline{FEM}]_i = [T]_i^T [FEM]_i \tag{2.21}$$

Consider a beam arbitrarily oriented, as shown in Figure 2.11. The degrees-of-freedom of the member are marked in both the local and global axes systems.

Now, the fixed end moments in the local and global axes are written as follows:

$$\{FEM\} = \begin{Bmatrix} M_p \\ M_q \\ V_r \\ V_s \\ H_t \\ H_h \end{Bmatrix}, \quad \{\overline{FEM}\} = \begin{Bmatrix} \overline{M}_p \\ \overline{M}_q \\ \overline{V}_r \\ \overline{V}_s \\ \overline{H}_t \\ \overline{H}_h \end{Bmatrix}$$

Therefore, the fixed end moment of global axes is a transformed value of the fixed end moment of local axes. Then, the joint load vector is obtained by reversing the sign of the global fixed end moment of every member. Finally, the joint load vector for the complete structure is obtained. Now, the joint load vector for the unrestrained degree can be obtained from the global joint load vector.

Step 7: Calculation of end moments and end shear

Then, calculate the unrestrained displacements by using the following equation:

$$\{\overline{\delta}_u\} = [\overline{K}_{uu}]^{-1} \{\overline{J}_L\}_u \tag{2.22}$$

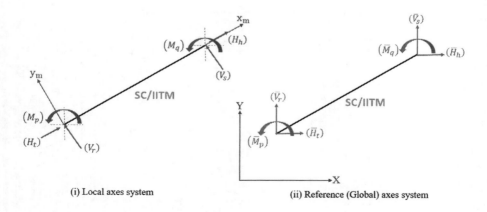

FIGURE 2.11 Degrees-of-freedom of beam in local and global axes system.

The end moments and shear of every member are obtained by the following equations:

$$\{M\}_i = [K]_i \{\delta_i\} + \{\text{FEM}\}_i$$
$$\{\bar{M}\}_i = [\bar{K}]_i \{\bar{\delta}_i\} + \{\bar{\text{FEM}}\}_i$$

(2.23)

Example problems with computer program

EXAMPLE 2.1:

Analyze the planar non-orthogonal structure shown in Figure 2.12 using the stiffness method. Given, $I = 0.0016$ m^4, $A = 0.120$ m^2 and E is constant.

SOLUTION:

Initially, mark the local and global axes for the structure.
For the member AB, $\theta = \tan^{-1}(4/2) = 63.435°$.
For the member BC, $\theta = 0°$.

1. *Calculation of transformation matrix coefficients and global labels:*

    The unrestrained and restrained degrees-of-freedom are marked in the structure, similar to that of the orthogonal structure. The local axes system for the members and the global axes system are also marked, as shown in Figure 2.13.
    Unrestrained degrees-of-freedom: 3 (θ_1, δ_2, δ_3)
    Restrained degrees-of-freedom: 6 (θ_4, δ_5, δ_6, θ_7, δ_8, δ_9)

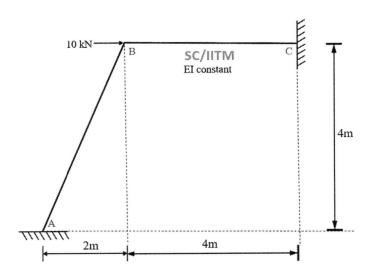

FIGURE 2.12 Non-orthogonal structure example.

Planar Non-Orthogonal Structures

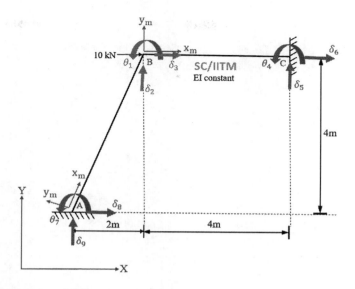

FIGURE 2.13 Degrees-of-freedom in local and global axes system.

Thus, the size of the total stiffness matrix will be 9×9, in which the submatrix for the unrestrained degrees-of-freedom will be of size 3×3.

Member Number	Ends j	Ends k	Length (m)	θ (Degrees)	C_x	C_y	Global Labels
1	A	B	4.472	+63.435	0.447	0.894	(7,1,9,2,8,3)
2	B	C	4	0	1	0	(1,4,2,5,3,6)

2. *Calculation of the local stiffness matrix:*

The stiffness matrix for the standard beam element including the axial deformation is given by,

$$[K]_i = \begin{bmatrix} \dfrac{4EI}{l} & \dfrac{2EI}{l} & \dfrac{6EI}{l^2} & -\dfrac{6EI}{l^2} & 0 & 0 \\ \dfrac{2EI}{l} & \dfrac{4EI}{l} & \dfrac{6EI}{l^2} & -\dfrac{6EI}{l^2} & 0 & 0 \\ \dfrac{6EI}{l^2} & \dfrac{6EI}{l^2} & \dfrac{12EI}{l^3} & -\dfrac{12EI}{l^3} & 0 & 0 \\ -\dfrac{6EI}{l^2} & -\dfrac{6EI}{l^2} & -\dfrac{12EI}{l^3} & \dfrac{12EI}{l^3} & 0 & 0 \\ 0 & 0 & 0 & 0 & \dfrac{AE}{l} & -\dfrac{AE}{l} \\ 0 & 0 & 0 & 0 & -\dfrac{AE}{l} & \dfrac{AE}{l} \end{bmatrix}$$

$$[K]_{AB} = E \begin{matrix} \circled{7} & \circled{1} & \circled{9} & \circled{2} & \circled{8} & \circled{3} \\ \begin{bmatrix} 0.0014 & 0.0007 & 0.0005 & -0.0005 & 0 & 0 \\ 0.0007 & 0.0014 & 0.0005 & -0.0005 & 0 & 0 \\ 0.0005 & 0.0005 & 0.0002 & -0.0002 & 0 & 0 \\ -0.0005 & -0.0005 & -0.0002 & 0.0002 & 0 & 0 \\ 0 & 0 & 0 & 0 & 0.0268 & -0.0268 \\ 0 & 0 & 0 & 0 & -0.0268 & 0.0268 \end{bmatrix} & \begin{matrix} \circled{7} \\ \circled{1} \\ \circled{9} \\ \circled{2} \\ \circled{8} \\ \circled{3} \end{matrix} \end{matrix}$$

$$[K]_{BC} = E \begin{matrix} \circled{1} & \circled{4} & \circled{2} & \circled{5} & \circled{3} & \circled{6} \\ \begin{bmatrix} 0.0016 & 0.0008 & 0.0006 & -0.0006 & 0 & 0 \\ 0.0008 & 0.0016 & 0.0006 & -0.0006 & 0 & 0 \\ 0.0006 & 0.0006 & 0.0003 & -0.0003 & 0 & 0 \\ -0.0006 & -0.0006 & -0.0003 & -0.0003 & 0 & 0 \\ 0 & 0 & 0 & 0 & 0.0300 & -0.0300 \\ 0 & 0 & 0 & 0 & -0.0300 & 0.0300 \end{bmatrix} & \begin{matrix} \circled{1} \\ \circled{4} \\ \circled{2} \\ \circled{5} \\ \circled{3} \\ \circled{6} \end{matrix} \end{matrix}$$

3. *Calculation of transformation matrix:*

   The transformation matrix for any member '*i*' is given by,

$$[T] = \begin{bmatrix} 1 & 0 & 0 & 0 & 0 & 0 \\ 0 & 1 & 0 & 0 & 0 & 0 \\ 0 & 0 & C_x & 0 & -C_y & 0 \\ 0 & 0 & 0 & C_x & 0 & -C_y \\ 0 & 0 & C_y & 0 & C_x & 0 \\ 0 & 0 & 0 & C_y & 0 & C_x \end{bmatrix}$$

$$[T]_{AB} = \begin{bmatrix} 1 & 0 & 0 & 0 & 0 & 0 \\ 0 & 1 & 0 & 0 & 0 & 0 \\ 0 & 0 & 0.4472 & 0 & -0.8944 & 0 \\ 0 & 0 & 0 & 0.4472 & 0 & -0.8944 \\ 0 & 0 & 0.0944 & 0 & 0.4472 & 0 \\ 0 & 0 & 0 & 0.8944 & 0 & 0.4472 \end{bmatrix}$$

$$[T]_{BC} = \begin{bmatrix} 1 & 0 & 0 & 0 & 0 & 0 \\ 0 & 1 & 0 & 0 & 0 & 0 \\ 0 & 0 & 1 & 0 & 0 & 0 \\ 0 & 0 & 0 & 1 & 0 & 0 \\ 0 & 0 & 0 & 0 & 1 & 0 \\ 0 & 0 & 0 & 0 & 0 & 1 \end{bmatrix}$$

Planar Non-Orthogonal Structures

4. *Estimation of the joint load vector:*

The structure does not have any member loading. There is a load acting only on the joint. Hence, fixed end moments will not be generated in the members. Thus, the joint load vector can be written directly as follows:

$$\{J_L\} = \begin{Bmatrix} 0 \\ 0 \\ +10 \\ \hline 0 \\ 0 \\ 0 \\ 0 \\ 0 \\ 0 \end{Bmatrix}_{9\times 1} \begin{matrix} ① \\ ② \\ ③ \\ ④ \\ ⑤ \\ ⑥ \\ ⑦ \\ ⑧ \\ ⑨ \end{matrix}$$

Thus, the joint load vector in unrestrained degrees-of-freedom is given by,

$$\{\bar{J}_{Lu}\} = \begin{Bmatrix} 0 \\ 0 \\ +10 \end{Bmatrix}_{3\times 1} \begin{matrix} ① \\ ② \\ ③ \end{matrix}$$

5. *Calculation of the global stiffness matrix:*

The stiffness matrix of every member, with respect to the global axes system, is obtained by the following equation:

$$\{\bar{K}\}_i = [T]_i^T [K]_i [T]_i$$

$$\bar{K}_{AB} = E \times 10^{-4} \begin{bmatrix} 14 & 7 & 2 & -2 & -4 & 4 \\ 7 & 14 & 2 & -2 & -4 & 4 \\ 2 & 2 & 215 & -215 & 106 & -106 \\ -2 & -2 & -215 & 215 & -106 & 106 \\ -4 & -4 & 106 & -106 & 55 & -55 \\ 4 & 4 & -106 & 106 & -55 & 55 \end{bmatrix} \begin{matrix} ⑦ \\ ① \\ ⑨ \\ ② \\ ⑧ \\ ③ \end{matrix}$$

with column labels ⑦ ① ⑨ ② ⑧ ③

$$\bar{K}_{BC} = E \times 10^{-4} \begin{bmatrix} 16 & 8 & 6 & -6 & 0 & 0 \\ 8 & 16 & 6 & -6 & 0 & 0 \\ 6 & 6 & 3 & -3 & 0 & 0 \\ -6 & -6 & -3 & 3 & 0 & 0 \\ 0 & 0 & 0 & 0 & 300 & -300 \\ 0 & 0 & 0 & 0 & -300 & 300 \end{bmatrix} \begin{matrix} ① \\ ④ \\ ② \\ ⑤ \\ ③ \\ ⑥ \end{matrix}$$

with column labels ① ④ ② ⑤ ③ ⑥

These matrices are then assembled to get the total stiffness matrix, which will be a 9×9 matrix with submatrices of unrestrained and restrained degrees-of-freedom.

$$\bar{K}_{TOTAL} = E \times 10^{-4} \begin{bmatrix} 38 & 4 & 4 & 8 & -6 & 0 & 7 & -4 & 2 \\ 4 & 218 & 106 & 6 & -3 & 0 & -2 & -106 & -215 \\ 4 & 106 & 355 & 0 & 0 & -300 & 4 & -55 & -106 \\ 8 & 6 & 0 & 16 & -6 & 0 & 0 & 0 & 0 \\ -6 & -3 & 0 & -6 & 3 & 0 & 0 & 0 & 0 \\ 0 & 0 & -300 & 0 & 0 & 300 & 0 & 0 & 0 \\ 7 & -2 & 4 & 0 & 0 & 0 & 14 & -4 & 2 \\ -4 & -106 & -55 & 0 & 0 & 0 & -4 & 55 & 106 \\ 2 & -215 & -106 & 0 & 0 & 0 & 2 & 106 & 215 \end{bmatrix} \begin{matrix} 1 \\ 2 \\ 3 \\ 4 \\ 5 \\ 6 \\ 7 \\ 8 \\ 9 \end{matrix}$$

$$\left[\bar{K}_{uu} \right] = E \times 10^{-4} \begin{bmatrix} 30 & 4 & 4 \\ 4 & 218 & 106 \\ 4 & 106 & 355 \end{bmatrix} \begin{matrix} 1 \\ 2 \\ 3 \end{matrix}$$

$$\left[\bar{K}_{uu} \right]^{-1} = \frac{1}{E} \begin{bmatrix} 330.863 & -4.561 & -2.631 \\ -4.561 & 53.769 & -16.055 \\ -2.631 & -16.055 & 32.980 \end{bmatrix}$$

Now, $\{\bar{\delta}_u\} = \left[\bar{K}_{uu} \right]^{-1} \{\bar{J}_L\}_u$

$$\{\bar{\delta}_u\} = \frac{1}{E} \begin{Bmatrix} -26.307 \\ -160.545 \\ 329.804 \end{Bmatrix}$$

$$\left[\bar{\delta}\right]_{AB} = \frac{1}{E} \begin{Bmatrix} 0 \\ -26.307 \\ 0 \\ -160.545 \\ 0 \\ 329.804 \end{Bmatrix}, \quad \left[\bar{\delta}\right]_{BC} = \frac{1}{E} \begin{Bmatrix} -26.307 \\ 0 \\ -160.545 \\ 0 \\ 329.804 \\ 0 \end{Bmatrix}$$

6. *Calculation of end moments and shear:*

$$\{\bar{M}\}_i = \left[\bar{K}\right]_i \{\bar{\delta}_i\} + \{\overline{FEM}\}_i$$

Planar Non-Orthogonal Structures

$$[\bar{M}]_{AB} = \begin{Bmatrix} \bar{M}_7 \\ \bar{M}_1 \\ \bar{V}_9 \\ \bar{V}_2 \\ \bar{H}_8 \\ \bar{H}_3 \end{Bmatrix} = \begin{Bmatrix} 0.1572 \\ 0.1384 \\ -0.0639 \\ 0.0639 \\ -0.1059 \\ 0.1059 \end{Bmatrix}$$

$$[\bar{M}]_{BC} = \begin{Bmatrix} \bar{M}_1 \\ \bar{M}_4 \\ \bar{V}_2 \\ \bar{V}_5 \\ \bar{H}_3 \\ \bar{H}_6 \end{Bmatrix} = \begin{Bmatrix} -0.1384 \\ -0.1174 \\ -0.0639 \\ 0.0639 \\ 9.8941 \\ -9.8941 \end{Bmatrix}$$

The member and final and moments are shown in Figures 2.14 and 2.15.

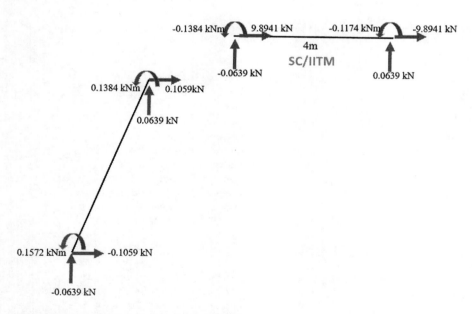

FIGURE 2.14 Member end moments and shear.

86 Advanced Structural Analysis with MATLAB®

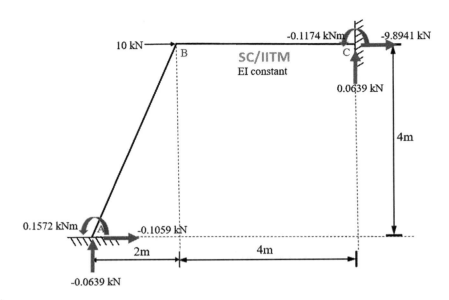

FIGURE 2.15 Final end moments and shear.

MATLAB® program:

```
%% stiffness matrix method
% Input
clc;
clear;
n = 2; % number of members
I = [0.0016 0.0016]; %Moment of inertis in m4
L = [4.472 4]; % length in m
A = [0.12 0.12]; % Area in m2
theta= [63.435 0]; % angle in degrees
uu = 3; % Number of unrestrained degrees of freedom
ur = 6; % Number of restrained degrees of freedom
uul = [1 2 3]; % global labels of unrestrained dof
url = [4 5 6 7 8 9]; % global labels of restrained dof
l1 = [7 1 9 2 8 3]; % Global labels for member 1
l2 = [1 4 2 5 3 6]; % Global labels for member 2
l= [l1; l2];
dof = uu + ur; % Degrees of freedom
Ktotal = zeros (dof);
Tt1 = zeros (6); % Transformation matrix for member 1
Tt2 = zeros (6); % Transformation matrix for member 2
Tt3 = zeros (6); % Transformation matrix for member 3
fem1= [0; 0; 0; 0; 0; 0]; % Local Fixed end moments of member 1
fem2= [0; 0; 0; 0; 0; 0]; % Local Fixed end moments of member 2

%% rotation coefficients for each member
rc1 = 4.*I./L;
```

Planar Non-Orthogonal Structures

```
rc2 = 2.*I./L;
rc3 = A./L;
cx = cosd(theta);
cy = sind(theta);

%% stiffness matrix 6 by 6
for i = 1:n
    Knew = zeros (dof);
    k1 = [rc1(i); rc2(i); (rc1(i)+rc2(i))/L(i);
(-(rc1(i)+rc2(i))/L(i)); 0; 0];
    k2 = [rc2(i); rc1(i); (rc1(i)+rc2(i))/L(i);
(-(rc1(i)+rc2(i))/L(i)); 0; 0;];
    k3 = [(rc1(i)+rc2(i))/L(i); (rc1(i)+rc2(i))/L(i);
(2*(rc1(i)+rc2(i))/(L(i)^2)); (-2*(rc1(i)+rc2(i))/(L(i)^2));
0; 0;];
    k4 = -k3;
    k5 = [0; 0; 0; 0; rc3(i); -rc3(i)];
    k6 = [0; 0; 0; 0; -rc3(i); rc3(i)];
    K = [k1 k2 k3 k4 k5 k6];
    fprintf ('Member Number =');
    disp (i);
    fprintf ('Local Stiffness matrix of member, [K] = \n');
    disp (K);
    T1 = [1; 0; 0; 0; 0; 0];
    T2 = [0; 1; 0; 0; 0; 0];
    T3 = [0; 0; cx(i); 0; cy(i); 0];
    T4 = [0; 0; 0; cx(i); 0; cy(i)];
    T5 = [0; 0; -cy(i); 0; cx(i); 0];
    T6 = [0; 0; 0; -cy(i); 0; cx(i)];
    T = [T1 T2 T3 T4 T5 T6];
    fprintf ('Tranformation matrix of member, [T] = \n');
    disp (T);
    Ttr = T';
    fprintf ('Tranformation matrix Transpose, [T] = \n');
    disp (Ttr);
    Kg = Ttr*K*T;
    fprintf ('Global Matrix, [K global] = \n');
    disp (Kg);
    for p = 1:6
        for q = 1:6
            Knew((l(i,p)),(l(i,q))) =Kg(p,q);
        end
    end
    Ktotal = Ktotal + Knew;
    if i == 1
        Tt1= T;
        Kg1=Kg;
        fembar1= Tt1'*fem1;
    elseif i == 2
        Tt2 = T;
```

```
            Kg2 = Kg;
            fembar2= Tt2'*fem2;
        end
end
fprintf ('Stiffness Matrix of complete structure, [Ktotal] = \n');
disp (Ktotal);
Kunr = zeros(3);
for x=1:uu
    for y=1:uu
        Kunr(x,y)= Ktotal(x,y);
    end
end
fprintf ('Unrestrained Stiffness sub-matrix, [Kuu] = \n');
disp (Kunr);
KuuInv= inv(Kunr);
fprintf ('Inverse of Unrestrained Stiffness sub-matrix,
[KuuInverse] = \n');
disp (KuuInv);

%% Creation of joint load vector
jl= [0; 0; 10; 0; 0; 0; 0; 0; 0]; % values given in kN or kNm
jlu = [0; 0; 10]; % load vector in unrestrained dof
delu = KuuInv*jlu;
fprintf ('Joint Load vector, [Jl] = \n');
disp (jl');
fprintf ('Unrestrained displacements, [DelU] = \n');
disp (delu');
delr = zeros (ur,1);
del = zeros (dof,1);
del = [delu; delr];
deli= zeros (6,1);
for i = 1:n
    for p = 1:6
        deli(p,1) = del((l(i,p)),1) ;
    end
    if i == 1
            delbar1 = deli;
            mbar1= (Kg1 * delbar1)+fembar1;
            fprintf ('Member Number =');
            disp (i);
            fprintf ('Global displacement matrix [DeltaBar] = \n');
            disp (delbar1');
            fprintf ('Global End moment matrix [MBar] = \n');
            disp (mbar1');
        elseif i == 2
            delbar2 = deli;
            mbar2= (Kg2 * delbar2)+fembar2;
            fprintf ('Member Number =');
            disp (i);
            fprintf ('Global displacement matrix [DeltaBar] = \n');
```

Planar Non-Orthogonal Structures

```
                disp (delbar2');
                fprintf ('Global End moment matrix [MBar] = \n');
                disp (mbar2');
        end
end

%% check
mbar = [mbar1'; mbar2'];
jf = zeros(dof,1);
for a=1:n
    for b=1:6 % size of k matrix
        d = l(a,b);
        jfnew = zeros(dof,1);
        jfnew(d,1)=mbar(a,b);
        jf=jf+jfnew;
    end
end
fprintf ('Joint forces = \n');
disp (jf');
```

MATLAB output:

```
Member Number = 1
Local Stiffness matrix of member, [K] =

    0.0014    0.0007    0.0005   -0.0005         0         0
    0.0007    0.0014    0.0005   -0.0005         0         0
    0.0005    0.0005    0.0002   -0.0002         0         0
   -0.0005   -0.0005   -0.0002    0.0002         0         0
         0         0         0         0    0.0268   -0.0268
         0         0         0         0   -0.0268    0.0268

Tranformation matrix of member, [T] =

    1.0000         0         0         0         0         0
         0    1.0000         0         0         0         0
         0         0    0.4472         0   -0.8944         0
         0         0         0    0.4472         0   -0.8944
         0         0    0.8944         0    0.4472         0
         0         0         0    0.8944         0    0.4472

Tranformation matrix Transpose, [T] =

    1.0000         0         0         0         0         0
         0    1.0000         0         0         0         0
         0         0    0.4472         0    0.8944         0
         0         0         0    0.4472         0    0.8944
         0         0   -0.8944         0    0.4472         0
         0         0         0   -0.8944         0    0.4472
```

Global Matrix, [K global] =

```
 0.0014    0.0007    0.0002   -0.0002   -0.0004    0.0004
 0.0007    0.0014    0.0002   -0.0002   -0.0004    0.0004
 0.0002    0.0002    0.0215   -0.0215    0.0106   -0.0106
-0.0002   -0.0002   -0.0215    0.0215   -0.0106    0.0106
-0.0004   -0.0004    0.0106   -0.0106    0.0055   -0.0055
 0.0004    0.0004   -0.0106    0.0106   -0.0055    0.0055
```

Member Number = 2
Local Stiffness matrix of member, [K] =

```
 0.0016    0.0008    0.0006   -0.0006        0         0
 0.0008    0.0016    0.0006   -0.0006        0         0
 0.0006    0.0006    0.0003   -0.0003        0         0
-0.0006   -0.0006   -0.0003    0.0003        0         0
      0         0         0         0    0.0300   -0.0300
      0         0         0         0   -0.0300    0.0300
```

Tranformation matrix of member, [T] =

```
1  0  0  0  0  0
0  1  0  0  0  0
0  0  1  0  0  0
0  0  0  1  0  0
0  0  0  0  1  0
0  0  0  0  0  1
```

Tranformation matrix Transpose, [T] =

```
1  0  0  0  0  0
0  1  0  0  0  0
0  0  1  0  0  0
0  0  0  1  0  0
0  0  0  0  1  0
0  0  0  0  0  1
```

Global Matrix, [K global] =

```
 0.0016    0.0008    0.0006   -0.0006        0         0
 0.0008    0.0016    0.0006   -0.0006        0         0
 0.0006    0.0006    0.0003   -0.0003        0         0
-0.0006   -0.0006   -0.0003    0.0003        0         0
      0         0         0         0    0.0300   -0.0300
      0         0         0         0   -0.0300    0.0300
```

Planar Non-Orthogonal Structures

```
Stiffness Matrix of complete structure, [Ktotal] =

   0.0030    0.0004    0.0004    0.0008   -0.0006        0    0.0007   -0.0004    0.0002
   0.0004    0.0218    0.0106    0.0006   -0.0003        0   -0.0002   -0.0106   -0.0215
   0.0004    0.0106    0.0355         0         0  -0.0300    0.0004   -0.0055   -0.0106
   0.0008    0.0006         0    0.0016   -0.0006        0         0         0         0
  -0.0006   -0.0003         0   -0.0006    0.0003        0         0         0         0
        0         0   -0.0300         0         0   0.0300         0         0         0
   0.0007   -0.0002    0.0004         0         0        0    0.0014   -0.0004    0.0002
  -0.0004   -0.0106   -0.0055         0         0        0   -0.0004    0.0055    0.0106
   0.0002   -0.0215   -0.0106         0         0        0    0.0002    0.0106    0.0215

Unrestrained Stiffness sub-matrix, [Kuu] =

   0.0030    0.0004    0.0004
   0.0004    0.0218    0.0106
   0.0004    0.0106    0.0355

Inverse of Unrestrained Stiffness sub-matrix, [KuuInverse] =

  330.8628   -4.5612   -2.6307
   -4.5612   53.7692  -16.0545
   -2.6307  -16.0545   32.9804

Joint Load vector, [Jl] =

   0    0    10    0    0    0    0    0    0

Unrestrained displacements, [DelU] =

  -26.3069  -160.5452   329.8036

Member Number =   1
Global displacement matrix [DeltaBar] =

   0   -26.3069    0   -160.5452    0    329.8036

Global End moment matrix [MBar] =

   0.1572    0.1384   -0.0639    0.0639   -0.1059    0.1059

Member Number =   2
Global displacement matrix [DeltaBar] =

  -26.3069    0   -160.5452    0    329.8036    0

Global End moment matrix [MBar] =

  -0.1384   -0.1174   -0.0639    0.0639    9.8941   -9.8941

Joint forces =

   0   0.0000   10.0000   -0.1174   0.0639   -9.8941   0.1572   -0.1059   -0.0639
```

EXAMPLE 2.2:

Analyze the planar non-orthogonal structure shown in Figure 2.16 using the stiffness method.

SOLUTION:

1. *Calculation of transformation matrix coefficients and global labels:*

   The unrestrained and restrained degrees-of-freedom are marked in the structure, similar to that of the orthogonal structure. The local axes system for the members and the global axes system are also marked, as shown in Figure 2.17.
   Unrestrained degrees-of-freedom: 6 ($\theta_1, \theta_2, \delta_3, \delta_4, \delta_5, \delta_6$)
   Restrained degrees-of-freedom: 6 ($\theta_7, \delta_8, \delta_9, \theta_{10}, \delta_{11}, \delta_{12}$)
   Thus, the size of the total stiffness matrix will be 12×12, in which the submatrix for the unrestrained degrees-of-freedom will be of size 6×6.

Member Number	Ends j	Ends k	Length (m)	θ (Degrees)	C_x	C_y	Global Labels
1	A	B	4	90	0	1	(7,1,9,4,8,3)
2	B	C	6	0	1	0	(1,2,4,6,3,5)
3	C	D	4.472	−63.435	0.447	−0.894	(2,10,6,12,5,11)

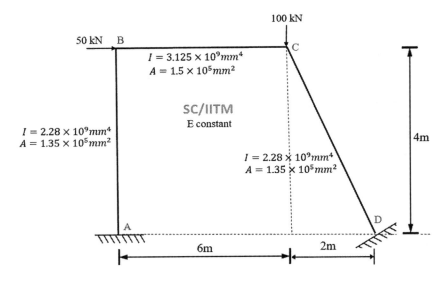

FIGURE 2.16 Non-orthogonal structure example.

Planar Non-Orthogonal Structures

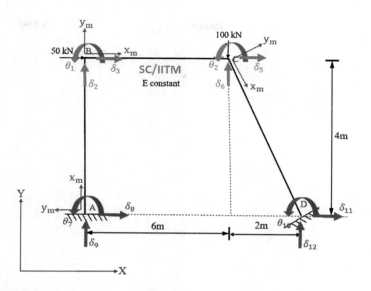

FIGURE 2.17 Degrees-of-freedom, local and global axes system.

2. *Calculation of the local stiffness matrix:*

The stiffness matrix for the standard beam element including the axial deformation is given by,

$$[K]_i = \begin{bmatrix} \dfrac{4EI}{l} & \dfrac{2EI}{l} & \dfrac{6EI}{l^2} & -\dfrac{6EI}{l^2} & 0 & 0 \\ \dfrac{2EI}{l} & \dfrac{4EI}{l} & \dfrac{6EI}{l^2} & -\dfrac{6EI}{l^2} & 0 & 0 \\ \dfrac{6EI}{l^2} & \dfrac{6EI}{l^2} & \dfrac{12EI}{l^3} & -\dfrac{12EI}{l^3} & 0 & 0 \\ -\dfrac{6EI}{l^2} & -\dfrac{6EI}{l^2} & -\dfrac{12EI}{l^3} & \dfrac{12EI}{l^3} & 0 & 0 \\ 0 & 0 & 0 & 0 & \dfrac{AE}{l} & -\dfrac{AE}{l} \\ 0 & 0 & 0 & 0 & -\dfrac{AE}{l} & \dfrac{AE}{l} \end{bmatrix}$$

$$[K]_{AB} = E \times 10^{-4} \begin{bmatrix} 23 & 11 & 9 & -9 & 0 & 0 \\ 11 & 23 & 9 & -9 & 0 & 0 \\ 9 & 9 & 4 & -4 & 0 & 0 \\ -9 & -9 & -4 & 4 & 0 & 0 \\ 0 & 0 & 0 & 0 & 338 & -338 \\ 0 & 0 & 0 & 0 & -338 & 338 \end{bmatrix} \begin{matrix} ⑦ \\ ① \\ ⑨ \\ ④ \\ ⑧ \\ ③ \end{matrix}$$

columns: ⑦ ① ⑨ ④ ⑧ ③

$$[K]_{BC} = E \times 10^{-4} \begin{bmatrix} 21 & 10 & 5 & -5 & 0 & 0 \\ 10 & 21 & 5 & -5 & 0 & 0 \\ 5 & 5 & 2 & -2 & 0 & 0 \\ -5 & -5 & -2 & 2 & 0 & 0 \\ 0 & 0 & 0 & 0 & 250 & -250 \\ 0 & 0 & 0 & 0 & -250 & 250 \end{bmatrix} \begin{matrix} (1) \\ (2) \\ (4) \\ (6) \\ (3) \\ (5) \end{matrix}$$

with column labels (1) (2) (4) (6) (3) (5)

$$[K]_{CD} = E \times 10^{-4} \begin{bmatrix} 20 & 10 & 7 & -7 & 0 & 0 \\ 10 & 20 & 7 & -7 & 0 & 0 \\ 7 & 7 & 3 & -3 & 0 & 0 \\ -7 & -7 & -3 & 3 & 0 & 0 \\ 0 & 0 & 0 & 0 & 302 & -302 \\ 0 & 0 & 0 & 0 & -302 & 302 \end{bmatrix} \begin{matrix} (2) \\ (10) \\ (6) \\ (11) \\ (5) \\ (12) \end{matrix}$$

with column labels (2) (10) (6) (11) (5) (12)

3. *Calculation of transformation matrix:*

The transformation matrix for any member '*i*' is given by,

$$[T] = \begin{bmatrix} 1 & 0 & 0 & 0 & 0 & 0 \\ 0 & 1 & 0 & 0 & 0 & 0 \\ 0 & 0 & C_x & 0 & -C_y & 0 \\ 0 & 0 & 0 & C_x & 0 & -C_y \\ 0 & 0 & C_y & 0 & C_x & 0 \\ 0 & 0 & 0 & C_y & 0 & C_x \end{bmatrix}$$

$$[T]_{AB} = \begin{bmatrix} 1 & 0 & 0 & 0 & 0 & 0 \\ 0 & 1 & 0 & 0 & 0 & 0 \\ 0 & 0 & 0 & 0 & -1 & 0 \\ 0 & 0 & 0 & 0 & 0 & -1 \\ 0 & 0 & 1 & 0 & 0 & 0 \\ 0 & 0 & 0 & 1 & 0 & 0 \end{bmatrix}$$

$$[T]_{BC} = \begin{bmatrix} 1 & 0 & 0 & 0 & 0 & 0 \\ 0 & 1 & 0 & 0 & 0 & 0 \\ 0 & 0 & 1 & 0 & 0 & 0 \\ 0 & 0 & 0 & 1 & 0 & 0 \\ 0 & 0 & 0 & 0 & 1 & 0 \\ 0 & 0 & 0 & 0 & 0 & 1 \end{bmatrix}$$

Planar Non-Orthogonal Structures

$$[T]_{CD} = \begin{bmatrix} 1 & 0 & 0 & 0 & 0 & 0 \\ 0 & 1 & 0 & 0 & 0 & 0 \\ 0 & 0 & 0.4472 & 0 & 0.8944 & 0 \\ 0 & 0 & 0 & 0.4472 & 0 & 0.8944 \\ 0 & 0 & -0.8944 & 0 & 0.4472 & 0 \\ 0 & 0 & 0 & 0.8944 & 0 & 0.4472 \end{bmatrix}$$

4. *Estimation of the joint load vector:*

The structure does not have any member loading. There is a load acting only on the joint. Hence, fixed end moments will not be generated in the members. Thus, the joint load vector can be written directly as follows:

$$\{J_L\} = \begin{Bmatrix} 0 \\ 0 \\ 50 \\ 0 \\ 0 \\ -100 \\ 0 \\ 0 \\ 0 \\ 0 \\ 0 \\ 0 \end{Bmatrix}_{12 \times 1} \begin{matrix} ① \\ ② \\ ③ \\ ④ \\ ⑤ \\ ⑥ \\ ⑦ \\ ⑧ \\ ⑧ \\ ⑩ \\ ⑪ \\ ⑫ \end{matrix}$$

Thus, the joint load vector in unrestrained degrees-of-freedom is given by,

$$\{\bar{J}_{Lu}\} = \begin{Bmatrix} 0 \\ 0 \\ 50 \\ 0 \\ 0 \\ -100 \end{Bmatrix}_{6 \times 1}$$

5. *Calculation of the global stiffness matrix:*

The stiffness matrix of every member, with respect to the global axes system, is obtained by the following equation:

$$\{\bar{K}\}_i = [T]_i^T [K]_i [T]_i$$

96   Advanced Structural Analysis with MATLAB®

$$\bar{K}_{AB} = E \times 10^{-4} \begin{array}{c} \;\;\;⑦\;\;\;①\;\;\;⑨\;\;\;④\;\;\;⑧\;\;\;③ \\ \begin{bmatrix} 23 & 11 & 0 & 0 & -9 & 9 \\ 11 & 23 & 0 & 0 & -9 & 9 \\ 0 & 0 & 338 & -338 & 0 & 0 \\ 0 & 0 & -338 & 338 & 0 & 0 \\ -9 & -9 & 0 & 0 & 4 & -4 \\ 9 & 9 & 0 & 0 & -4 & 4 \end{bmatrix} \begin{array}{c} ⑦ \\ ① \\ ⑨ \\ ④ \\ ⑧ \\ ③ \end{array} \end{array}$$

$$\bar{K}_{BC} = E \times 10^{-4} \begin{array}{c} \;\;\;①\;\;\;②\;\;\;④\;\;\;⑥\;\;\;③\;\;\;⑤ \\ \begin{bmatrix} 21 & 10 & 5 & -5 & 0 & 0 \\ 10 & 21 & 5 & -5 & 0 & 0 \\ 5 & 5 & 2 & -2 & 0 & 0 \\ -5 & -5 & -2 & 2 & 0 & 0 \\ 0 & 0 & 0 & 0 & 250 & -250 \\ 0 & 0 & 0 & 0 & -250 & 250 \end{bmatrix} \begin{array}{c} ① \\ ② \\ ④ \\ ⑥ \\ ③ \\ ⑤ \end{array} \end{array}$$

$$\bar{K}_{CD} = E \times 10^{-4} \begin{array}{c} \;\;\;②\;\;\;⑩\;\;\;⑥\;\;\;⑪\;\;\;⑤\;\;\;⑫ \\ \begin{bmatrix} 20 & 10 & 3 & -3 & 6 & -6 \\ 10 & 20 & 3 & -3 & 6 & -6 \\ 3 & 3 & 242 & -242 & -120 & 120 \\ -3 & -3 & -242 & 242 & 120 & -120 \\ 6 & 6 & -120 & 120 & 63 & -63 \\ -6 & -6 & 120 & 120 & -63 & 63 \end{bmatrix} \begin{array}{c} ② \\ ⑩ \\ ⑥ \\ ⑪ \\ ⑤ \\ ⑫ \end{array} \end{array}$$

$$\bar{K}_{\text{TOTAL}} = E \times 10^{-4} \begin{array}{c} \;①\;\;②\;\;③\;\;④\;\;⑤\;\;⑥\;\;⑦\;\;⑧\;\;⑨\;\;⑩\;\;⑪\;\;⑫ \\ \left[\begin{array}{cccccc|cccccc} 44 & 10 & 9 & 5 & 0 & -5 & 11 & -9 & 0 & 0 & 0 & 0 \\ 10 & 41 & 0 & 5 & 6 & -2 & 0 & 0 & 0 & 10 & -6 & -3 \\ 9 & 0 & 254 & 0 & -250 & 0 & 9 & -4 & 0 & 0 & 0 & 0 \\ 5 & 5 & 0 & 339 & 0 & -2 & 0 & 0 & -338 & 0 & 0 & 0 \\ 0 & 6 & -250 & 0 & 313 & -120 & 0 & 0 & 0 & 6 & -63 & 120 \\ -5 & -2 & 0 & -2 & -120 & 244 & 0 & 0 & 0 & 3 & 120 & -242 \\ \hline 00 & 0 & 9 & 0 & 0 & 0 & 23 & 23 & 0 & 0 & 0 & 0 \\ -9 & 0 & -4 & 0 & 0 & 0 & -9 & -9 & 0 & 0 & 0 & 0 \\ 0 & 0 & 0 & 338 & 0 & 0 & 0 & 0 & 338 & 0 & 0 & 0 \\ 0 & 10 & 0 & 0 & 6 & 3 & 0 & 0 & 0 & 20 & -6 & -3 \\ 0 & -6 & 0 & 0 & -63 & 120 & 0 & 0 & 0 & -6 & 63 & -120 \\ 0 & -3 & 0 & 0 & 120 & -242 & 0 & 0 & 0 & -3 & -120 & 242 \end{array} \right] \begin{array}{c} ① \\ ② \\ ③ \\ ④ \\ ⑤ \\ ⑥ \\ ⑦ \\ ⑧ \\ ⑨ \\ ⑩ \\ ⑪ \\ ⑫ \end{array} \end{array}$$

Planar Non-Orthogonal Structures

$$[\bar{K}_{uu}] = E \times 10^{-4} \begin{array}{c} \\ \end{array} \begin{array}{cccccc} \textcircled{1} & \textcircled{2} & \textcircled{3} & \textcircled{4} & \textcircled{5} & \textcircled{6} \end{array}$$

$$[\bar{K}_{uu}] = E \times 10^{-4} \begin{bmatrix} 44 & 10 & 9 & 5 & 0 & -5 \\ 10 & 41 & 0 & 5 & 6 & -2 \\ 9 & 0 & 254 & 0 & -250 & 0 \\ 5 & 5 & 0 & 339 & 0 & -2 \\ 0 & 6 & -250 & 0 & 313 & -120 \\ -5 & -2 & 0 & -2 & -120 & 244 \end{bmatrix} \begin{array}{c} \textcircled{1} \\ \textcircled{2} \\ \textcircled{3} \\ \textcircled{4} \\ \textcircled{5} \\ \textcircled{6} \end{array}$$

$$[\bar{K}_{uu}]^{-1} = \frac{1}{E} \begin{bmatrix} 264 & -46.9 & -161.7 & -3.7 & -155.4 & -71 \\ -46.9 & 271 & -126.7 & -3.8 & -130.5 & -62.6 \\ -161.7 & -126.7 & 1376.3 & 7.8 & 1354.3 & 659.3 \\ -3.7 & -3.8 & 7.8 & 29.6 & 7.8 & 3.9 \\ -155.4 & -130.5 & 1354.3 & 7.8 & 1372.1 & 668.1 \\ -71 & -62.6 & 659.3 & 3.9 & 668.1 & 366.5 \end{bmatrix}$$

Now, $\{\bar{\delta}_u\} = [\bar{K}_{uu}]^{-1} \{\bar{J}_L\}_u$

$$\{\bar{\delta}_u\} = \frac{1}{E} \begin{Bmatrix} -986 \\ -75.8 \\ 2882.9 \\ -2.5 \\ 898.5 \\ -3682.2 \end{Bmatrix}$$

$$[\bar{\delta}]_{AB} = \frac{1}{E} \begin{Bmatrix} 0 \\ -986 \\ 0 \\ -2.5 \\ 0 \\ 2882.9 \end{Bmatrix}, \quad [\bar{\delta}]_{BC} = \frac{1}{E} \begin{Bmatrix} -986 \\ -75.8 \\ -2.5 \\ -3682.2 \\ 2882.9 \\ 898.5 \end{Bmatrix}, \quad [\bar{\delta}]_{CD} = \frac{1}{E} \begin{Bmatrix} -75.8 \\ 0 \\ -3682.2 \\ 0 \\ 898.5 \\ 0 \end{Bmatrix}$$

6. Calculation of end moments and shear:

$$\{\bar{M}\}_i = [\bar{K}]_i \{\bar{\delta}_i\} + \{\bar{FEM}\}_i$$

$$[\bar{M}]_{AB} = \begin{Bmatrix} \bar{M}_7 \\ \bar{M}_1 \\ \bar{V}_9 \\ \bar{V}_4 \\ \bar{H}_8 \\ \bar{H}_3 \end{Bmatrix} = \begin{Bmatrix} 1.3408 \\ 0.2167 \\ 0.0858 \\ -0.0858 \\ -0.3894 \\ 0.3894 \end{Bmatrix}$$

98   Advanced Structural Analysis with MATLAB®

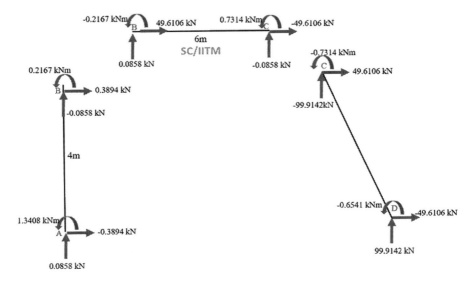

FIGURE 2.18   Member end moments and shear.

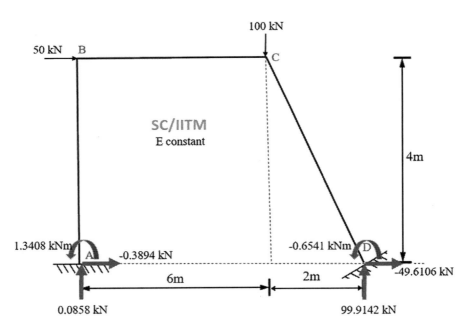

FIGURE 2.19   Final end moments and shear.

Planar Non-Orthogonal Structures

$$[\bar{M}]_{BC} = \begin{Bmatrix} \bar{M}_1 \\ \bar{M}_2 \\ \bar{V}_4 \\ \bar{V}_6 \\ \bar{H}_3 \\ \bar{H}_5 \end{Bmatrix} = \begin{Bmatrix} -0.2167 \\ 0.7314 \\ 0.0858 \\ -0.0858 \\ 49.6106 \\ -49.6106 \end{Bmatrix}$$

$$[\bar{M}]_{CD} = \begin{Bmatrix} \bar{M}_2 \\ \bar{M}_{10} \\ \bar{V}_6 \\ \bar{V}_{12} \\ \bar{H}_5 \\ \bar{H}_{11} \end{Bmatrix} = \begin{Bmatrix} -0.7314 \\ -0.6541 \\ -99.9142 \\ 99.9142 \\ 49.6106 \\ -49.6106 \end{Bmatrix}$$

The member and final and moments are shown in Figures 2.18 and 2.19.

MATLAB program

```
%% stiffness matrix method
% Input
clc;
clear;
n = 3; % number of members
I = [0.00228 0.003125 0.00228]; %Moment of inertis in m4
L = [4 6 4.472]; % length in m
A = [0.135 0.15 0.135]; % Area in m2
theta= [90 0 -63.435]; % angle in degrees
uu = 6; % Number of unrestrained degrees of freedom
ur = 6; % Number of restrained degrees of freedom
uul = [1 2 3 4 5 6]; % global labels of unrestrained dof
url = [7 8 9 10 11 12]; % global labels of restrained dof
l1 = [7 1 9 4 8 3]; % Global labels for member 1
l2 = [1 2 4 6 3 5]; % Global labels for member 2
l3 = [2 10 6 12 5 11]; % Global labels for member 3
l= [l1; l2; l3];
dof = uu + ur; % Degrees of freedom
Ktotal = zeros (dof);
Tt1 = zeros (6); % Transformation matrix for member 1
Tt2 = zeros (6); % Transformation matrix for member 2
Tt3 = zeros (6); % Transformation matrix for member 3
fem1= [0; 0; 0; 0; 0; 0]; % Local Fixed end moments of member 1
fem2= [0; 0; 0; 0; 0; 0]; % Local Fixed end moments of member 2
fem3= [0; 0; 0; 0; 0; 0]; % Local Fixed end moments of member 3
```

```matlab
%% rotation coefficients for each member
rc1 = 4.*I./L;
rc2 = 2.*I./L;
rc3 = A./L;
cx = cosd(theta);
cy = sind(theta);

%% stiffness matrix 6 by 6
for i = 1:n
    Knew = zeros (dof);
    k1 = [rc1(i); rc2(i); (rc1(i)+rc2(i))/L(i); (-(rc1(i)+rc2(i))/L(i)); 0; 0];
    k2 = [rc2(i); rc1(i); (rc1(i)+rc2(i))/L(i); (-(rc1(i)+rc2(i))/L(i)); 0; 0;];
    k3 = [(rc1(i)+rc2(i))/L(i); (rc1(i)+rc2(i))/L(i); (2*(rc1(i)+rc2(i))/(L(i)^2)); (-2*(rc1(i)+rc2(i))/(L(i)^2)); 0; 0;];
    k4 = -k3;
    k5 = [0; 0; 0; 0; rc3(i); -rc3(i)];
    k6 = [0; 0; 0; 0; -rc3(i); rc3(i)];
    K = [k1 k2 k3 k4 k5 k6];
    fprintf ('Member Number =');
    disp (i);
    fprintf ('Local Stiffness matrix of member, [K] = \n');
    disp (K);
    T1 = [1; 0; 0; 0; 0; 0];
    T2 = [0; 1; 0; 0; 0; 0];
    T3 = [0; 0; cx(i); 0; cy(i); 0];
    T4 = [0; 0; 0; cx(i); 0; cy(i)];
    T5 = [0; 0; -cy(i); 0; cx(i); 0];
    T6 = [0; 0; 0; -cy(i); 0; cx(i)];
    T = [T1 T2 T3 T4 T5 T6];
    fprintf ('Tranformation matrix of member, [T] = \n');
    disp (T);
    Ttr = T';
    fprintf ('Tranformation matrix Transpose, [T] = \n');
    disp (Ttr);
    Kg = Ttr*K*T;
    fprintf ('Global Matrix, [K global] = \n');
    disp (Kg);
    for p = 1:6
        for q = 1:6
            Knew((l(i,p)),(l(i,q))) =Kg(p,q);
        end
    end
    Ktotal = Ktotal + Knew;
    if i == 1
        Tt1= T;
        Kg1=Kg;
        fembar1= Tt1'*fem1;
    elseif i == 2
```

Planar Non-Orthogonal Structures

```
            Tt2 = T;
            Kg2 = Kg;
            fembar2= Tt2'*fem2;
        else
            Tt3 = T;
            Kg3=Kg;
            fembar3= Tt3'*fem3;
        end
end
fprintf ('Stiffness Matrix of complete structure, [Ktotal] = \n');
disp (Ktotal);
Kunr = zeros(6);
for x=1:uu
    for y=1:uu
        Kunr(x,y)= Ktotal(x,y);
    end
end
fprintf ('Unrestrained Stiffness sub-matix, [Kuu] = \n');
disp (Kunr);
KuuInv= inv(Kunr);
fprintf ('Inverse of Unrestrained Stiffness sub-matrix, [KuuInverse] = \n');
disp (KuuInv);

%% Creation of joint load vector
jl= [0; 0; 50; 0; 0; -100; 0; 0; 0; 0; 0; 0]; % values given in kN or kNm
jlu = jl(1:uu,1); % load vector in unrestrained dof
delu = KuuInv*jlu;
fprintf ('Joint Load vector, [Jl] = \n');
disp (jl');
fprintf ('Unrestrained displacements, [DelU] = \n');
disp (delu');
delr = zeros (ur,1);
del = zeros (dof,1);
del = [delu; delr];
deli= zeros (6,1);
for i = 1:n
    for p = 1:6
        deli(p,1) = del((l(i,p)),1) ;
    end
    if i == 1
            delbar1 = deli;
            mbar1= (Kg1 * delbar1)+fembar1;
            fprintf ('Member Number =');
            disp (i);
            fprintf ('Global displacement matrix [DeltaBar] = \n');
            disp (delbar1');
            fprintf ('Global End moment matrix [MBar] = \n');
            disp (mbar1');
```

```
            elseif i == 2
            delbar2 = deli;
            mbar2= (Kg2 * delbar2)+fembar2;
            fprintf ('Member Number =');
            disp (i);
            fprintf ('Global displacement matrix [DeltaBar] = \n');
            disp (delbar2');
            fprintf ('Global End moment matrix [MBar] = \n');
            disp (mbar2');
        else
            delbar3 = deli;
            mbar3= (Kg3 * delbar3)+fembar3;
            fprintf ('Member Number =');
            disp (i);
            fprintf ('Global displacement matrix [DeltaBar] = \n');
            disp (delbar3');
            fprintf ('Global End moment matrix [MBar] = \n');
            disp (mbar3');
        end
end

%% check
mbar = [mbar1'; mbar2'; mbar3'];
jf = zeros(dof,1);
for a=1:n
    for b=1:6 % size of k matrix
        d = l(a,b);
        jfnew = zeros(dof,1);
        jfnew(d,1)=mbar(a,b);
        jf=jf+jfnew;
    end
end
fprintf ('Joint forces = \n');
disp (jf');
```

MATLAB output:

Member Number = 1
Local Stiffness matrix of member, [K] =

0.0023	0.0011	0.0009	-0.0009	0	0
0.0011	0.0023	0.0009	-0.0009	0	0
0.0009	0.0009	0.0004	-0.0004	0	0
-0.0009	-0.0009	-0.0004	0.0004	0	0
0	0	0	0	0.0338	-0.0338
0	0	0	0	-0.0338	0.0338

Planar Non-Orthogonal Structures

Tranformation matrix of member, [T] =

```
1  0  0  0   0   0
0  1  0  0   0   0
0  0  0  0  -1   0
0  0  0  0   0  -1
0  0  1  0   0   0
0  0  0  1   0   0
```

Tranformation matrix Transpose, [T] =

```
1  0   0   0   0  0
0  1   0   0   0  0
0  0   0   0   1  0
0  0   0   0   0  1
0  0  -1   0   0  0
0  0   0  -1   0  0
```

Global Matrix, [K global] =

```
 0.0023   0.0011    0        0       -0.0009   0.0009
 0.0011   0.0023    0        0       -0.0009   0.0009
 0        0         0.0338  -0.0338   0        0
 0        0        -0.0338   0.0338   0        0
-0.0009  -0.0009    0        0        0.0004  -0.0004
 0.0009   0.0009    0        0       -0.0004   0.0004
```

Member Number = 2
Local Stiffness matrix of member, [K] =

```
 0.0021   0.0010   0.0005  -0.0005   0        0
 0.0010   0.0021   0.0005  -0.0005   0        0
 0.0005   0.0005   0.0002  -0.0002   0        0
-0.0005  -0.0005  -0.0002   0.0002   0        0
 0        0        0        0        0.0250  -0.0250
 0        0        0        0       -0.0250   0.0250
```

Tranformation matrix of member, [T] =

```
1  0  0  0  0  0
0  1  0  0  0  0
0  0  1  0  0  0
0  0  0  1  0  0
0  0  0  0  1  0
0  0  0  0  0  1
```

Tranformation matrix Transpose, [T] =

```
1  0  0  0  0  0
0  1  0  0  0  0
0  0  1  0  0  0
0  0  0  1  0  0
0  0  0  0  1  0
0  0  0  0  0  1
```

Global Matrix, [K global] =

```
 0.0021   0.0010   0.0005  -0.0005        0        0
 0.0010   0.0021   0.0005  -0.0005        0        0
 0.0005   0.0005   0.0002  -0.0002        0        0
-0.0005  -0.0005  -0.0002   0.0002        0        0
      0        0        0        0   0.0250  -0.0250
      0        0        0        0  -0.0250   0.0250
```

Member Number = 3
Local Stiffness matrix of member, [K] =

```
 0.0020   0.0010   0.0007  -0.0007        0        0
 0.0010   0.0020   0.0007  -0.0007        0        0
 0.0007   0.0007   0.0003  -0.0003        0        0
-0.0007  -0.0007  -0.0003   0.0003        0        0
      0        0        0        0   0.0302  -0.0302
      0        0        0        0  -0.0302   0.0302
```

Tranformation matrix of member, [T] =

```
1.0000        0        0        0        0        0
     0   1.0000        0        0        0        0
     0        0   0.4472        0   0.8944        0
     0        0        0   0.4472        0   0.8944
     0        0  -0.8944        0   0.4472        0
     0        0        0  -0.8944        0   0.4472
```

Tranformation matrix Transpose, [T] =

```
1.0000        0        0        0        0        0
     0   1.0000        0        0        0        0
     0        0   0.4472        0  -0.8944        0
     0        0        0   0.4472        0  -0.8944
     0        0   0.8944        0   0.4472        0
     0        0        0   0.8944        0   0.4472
```

Planar Non-Orthogonal Structures

Global Matrix, [K global] =

```
  0.0020   0.0010   0.0003  -0.0003   0.0006  -0.0006
  0.0010   0.0020   0.0003  -0.0003   0.0006  -0.0006
  0.0003   0.0003   0.0242  -0.0242  -0.0120   0.0120
 -0.0003  -0.0003  -0.0242   0.0242   0.0120  -0.0120
  0.0006   0.0006  -0.0120   0.0120   0.0063  -0.0063
 -0.0006  -0.0006   0.0120  -0.0120  -0.0063   0.0063
```

Stiffness Matrix of complete structure, [Ktotal] =
Columns 1 through 10

```
  0.0044   0.0010   0.0009   0.0005        0  -0.0005   0.0011  -0.0009        0        0
  0.0010   0.0041        0   0.0005   0.0006  -0.0002        0        0        0   0.0010
  0.0009        0   0.0254        0  -0.0250        0   0.0009  -0.0004        0        0
  0.0005   0.0005        0   0.0339        0  -0.0002        0        0  -0.0338        0
       0   0.0006  -0.0250        0   0.0313  -0.0120        0        0        0   0.0006
 -0.0005  -0.0002        0  -0.0002  -0.0120   0.0244        0        0        0   0.0003
  0.0011        0   0.0009        0        0        0   0.0023  -0.0009        0        0
 -0.0009        0  -0.0004        0        0        0  -0.0009   0.0004        0        0
       0        0        0  -0.0338        0        0        0        0   0.0338        0
       0   0.0010        0        0   0.0006   0.0003        0        0        0   0.0020
       0  -0.0006        0        0  -0.0063   0.0120        0        0        0  -0.0006
       0  -0.0003        0        0   0.0120  -0.0242        0        0        0  -0.0003
```

Columns 11 through 12

```
       0        0
 -0.0006  -0.0003
       0        0
       0        0
 -0.0063   0.0120
  0.0120  -0.0242
       0        0
       0        0
       0        0
 -0.0006  -0.0003
  0.0063  -0.0120
 -0.0120   0.0242
```

Unrestrained Stiffness sub-matrix, [Kuu] =

```
  0.0044   0.0010   0.0009   0.0005        0  -0.0005
  0.0010   0.0041        0   0.0005   0.0006  -0.0002
  0.0009        0   0.0254        0  -0.0250        0
  0.0005   0.0005        0   0.0339        0  -0.0002
       0   0.0006  -0.0250        0   0.0313  -0.0120
 -0.0005  -0.0002        0  -0.0002  -0.0120   0.0244
```

Inverse of Unrestrained Stiffness sub-matrix, [KuuInverse] =

```
1.0e+03 *
  0.2640   -0.0469   -0.1617   -0.0037   -0.1554   -0.0710
 -0.0469    0.2710   -0.1267   -0.0038   -0.1305   -0.0626
 -0.1617   -0.1267    1.3763    0.0078    1.3543    0.6593
 -0.0037   -0.0038    0.0078    0.0296    0.0078    0.0039
 -0.1554   -0.1305    1.3543    0.0078    1.3721    0.6681
 -0.0710   -0.0626    0.6593    0.0039    0.6681    0.3665
```

Joint Load vector, [Jl] =

```
  0   0   50   0   0   -100   0   0   0   0   0   0
```

Unrestrained displacements, [DelU] =

```
1.0e+03 *
 -0.9860   -0.0758   2.8829   -0.0025   0.8985   -3.6822
```

Member Number = 1
Global displacement matrix [DeltaBar] =

```
1.0e+03 *
      0   -0.9860   0   -0.0025   0   2.8829
```

Global End moment matrix [MBar] =

```
 1.3408   0.2167   0.0858   -0.0858   -0.3894   0.3894
```

Member Number = 2
Global displacement matrix [DeltaBar] =

```
1.0e+03 *
 -0.9860   -0.0758   -0.0025   -3.6822   2.8829   0.8985
```

Global End moment matrix [MBar] =

```
 -0.2167   0.7314   0.0858   -0.0858   49.6106   -49.6106
```

Member Number =   3
Global displacement matrix [DeltaBar] =

```
1.0e+03 *
 -0.0758   0   -3.6822   0   0.8985   0
```

Global End moment matrix [MBar] =

```
 -0.7314   -0.6541   -99.9142   99.9142   49.6106   -49.6106
```

Joint forces =

```
    0.0000    -0.0000    50.0000   -0.0000   -0.0000
 -100.0000     1.3408    -0.3894    0.0858   -0.6541   -49.6106   99.9142
```

Planar Non-Orthogonal Structures

EXAMPLE 2.3:
Analyze the planar non-orthogonal structure shown in Figure 2.20 using the stiffness method.

SOLUTION:

1. *Calculation of transformation matrix coefficients and global labels:*

   The unrestrained and restrained degrees-of-freedom are marked in the structure, similar to that of the orthogonal structure. The local axes system for the members and the global axes system are also marked, as shown in Figure 2.21.
   Unrestrained degrees-of-freedom: 6 (θ_1, θ_2, δ_3, δ_4, δ_5, δ_6)
   Restrained degrees-of-freedom: 6 (θ_7, δ_8, δ_9, θ_{10}, δ_{11}, δ_{12})
   Thus, the size of the total stiffness matrix will be 12×12, in which the submatrix for the unrestrained degrees-of-freedom will be of size 6×6.

Member Number	Ends j	k	Length (m)	θ (Degrees)	C_x	C_y	Global Labels
1	A	B	4	90	0	1	(7,1,9,4,8,3)
2	B	C	6	0	1	0	(1,2,4,6,3,5)
3	C	D	5	−53.123	0.6	−0.8	(2,10,6,12,5,11)

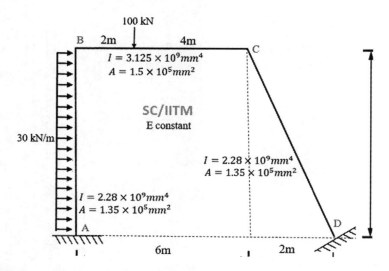

FIGURE 2.20 Non-orthogonal structure example.

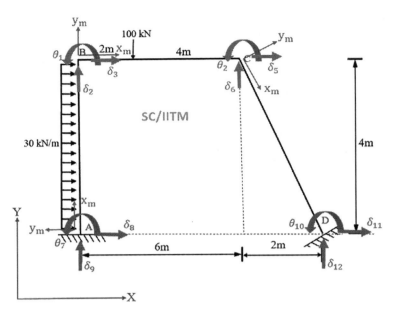

FIGURE 2.21 Degrees-of-freedom marked in local and global axes system.

2. *Calculation of the local stiffness matrix:*
   The stiffness matrix for the standard beam element, including the axial deformation, is given by,

$$[K]_i = \begin{bmatrix} \dfrac{4EI}{l} & \dfrac{2EI}{l} & \dfrac{6EI}{l^2} & -\dfrac{6EI}{l^2} & 0 & 0 \\ \dfrac{2EI}{l} & \dfrac{4EI}{l} & \dfrac{6EI}{l^2} & -\dfrac{6EI}{l^2} & 0 & 0 \\ \dfrac{6EI}{l^2} & \dfrac{6EI}{l^2} & \dfrac{12EI}{l^3} & -\dfrac{12EI}{l^3} & 0 & 0 \\ -\dfrac{6EI}{l^2} & -\dfrac{6EI}{l^2} & -\dfrac{12EI}{l^3} & \dfrac{12EI}{l^3} & 0 & 0 \\ 0 & 0 & 0 & 0 & \dfrac{AE}{l} & -\dfrac{AE}{l} \\ 0 & 0 & 0 & 0 & -\dfrac{AE}{l} & \dfrac{AE}{l} \end{bmatrix}$$

$$\begin{array}{cccccc} \small\textcircled{7} & \small\textcircled{1} & \small\textcircled{9} & \small\textcircled{4} & \small\textcircled{8} & \small\textcircled{3} \end{array}$$

$$[K]_{AB} = E \times 10^{-4} \begin{bmatrix} 23 & 11 & 9 & -9 & 0 & 0 \\ 11 & 23 & 9 & -9 & 0 & 0 \\ 9 & 9 & 4 & -4 & 0 & 0 \\ -9 & -9 & -4 & 4 & 0 & 0 \\ 0 & 0 & 0 & 0 & 338 & -338 \\ 0 & 0 & 0 & 0 & -338 & 338 \end{bmatrix} \begin{matrix} \small\textcircled{7} \\ \small\textcircled{1} \\ \small\textcircled{9} \\ \small\textcircled{4} \\ \small\textcircled{8} \\ \small\textcircled{3} \end{matrix}$$

Planar Non-Orthogonal Structures

$$[K]_{BC} = E \times 10^{-4} \begin{bmatrix} 21 & 10 & 5 & -5 & 0 & 0 \\ 10 & 21 & 5 & -5 & 0 & 0 \\ 5 & 5 & 2 & -2 & 0 & 0 \\ -5 & -5 & -2 & 2 & 0 & 0 \\ 0 & 0 & 0 & 0 & 250 & -250 \\ 0 & 0 & 0 & 0 & -250 & 250 \end{bmatrix} \begin{matrix} ①\\②\\④\\⑥\\③\\⑤ \end{matrix}$$

$\quad\quad\quad\quad\quad ① \;\; ② \;\; ④ \;\; ⑥ \;\; ③ \;\; ⑤$

$\quad\quad\quad\quad\quad ② \;\; ⑩ \;\; ⑥ \;\; ⑫ \;\; ⑤ \;\; ⑪$

$$[K]_{CD} = E \times 10^{-4} \begin{bmatrix} 18 & 9 & 5 & -5 & 0 & 0 \\ 9 & 18 & 5 & -5 & 0 & 0 \\ 5 & 5 & 2 & -2 & 0 & 0 \\ -5 & -5 & -2 & 2 & 0 & 0 \\ 0 & 0 & 0 & 0 & 270 & -270 \\ 0 & 0 & 0 & 0 & -270 & 270 \end{bmatrix} \begin{matrix} ②\\⑩\\⑥\\⑫\\⑤\\⑪ \end{matrix}$$

3. *Calculation of transformation matrix:*
   The transformation matrix for any member '*i*' is given by,

$$[T] = \begin{bmatrix} 1 & 0 & 0 & 0 & 0 & 0 \\ 0 & 1 & 0 & 0 & 0 & 0 \\ 0 & 0 & C_x & 0 & -C_y & 0 \\ 0 & 0 & 0 & C_x & 0 & -C_y \\ 0 & 0 & C_y & 0 & C_x & 0 \\ 0 & 0 & 0 & C_y & 0 & C_x \end{bmatrix}$$

$$[T]_{AB} = \begin{bmatrix} 1 & 0 & 0 & 0 & 0 & 0 \\ 0 & 1 & 0 & 0 & 0 & 0 \\ 0 & 0 & 0 & 0 & -1 & 0 \\ 0 & 0 & 0 & 0 & 0 & -1 \\ 0 & 0 & 1 & 0 & 0 & 0 \\ 0 & 0 & 0 & 1 & 0 & 0 \end{bmatrix}$$

$$[T]_{BC} = \begin{bmatrix} 1 & 0 & 0 & 0 & 0 & 0 \\ 0 & 1 & 0 & 0 & 0 & 0 \\ 0 & 0 & 1 & 0 & 0 & 0 \\ 0 & 0 & 0 & 1 & 0 & 0 \\ 0 & 0 & 0 & 0 & 1 & 0 \\ 0 & 0 & 0 & 0 & 0 & 1 \end{bmatrix}$$

$$[T]_{CD} = \begin{bmatrix} 1 & 0 & 0 & 0 & 0 & 0 \\ 0 & 1 & 0 & 0 & 0 & 0 \\ 0 & 0 & 0.6001 & 0 & 0.7999 & 0 \\ 0 & 0 & 0 & 0.6001 & 0 & 0.7999 \\ 0 & 0 & -0.7999 & 0 & 0.6001 & 0 \\ 0 & 0 & 0 & -0.7999 & 0 & 0.6001 \end{bmatrix}$$

4. *Estimation of joint load vector:*
   The fixed end moments on the members are calculated as follows:
   Member AB:

$$\{FEM\}_{AB} = \begin{Bmatrix} 40 \\ -40 \\ 60 \\ 60 \\ 0 \\ 0 \end{Bmatrix}, \quad \{F\bar{E}M\}_{AB} = \begin{Bmatrix} 40 \\ -40 \\ 0 \\ 0 \\ -60 \\ -60 \end{Bmatrix}$$

Member BC:

$$\{FEM\}_{BC} = \begin{Bmatrix} 89.89 \\ -44.44 \\ 66.67 \\ 33.333 \\ 0 \\ 0 \end{Bmatrix}, \quad \{F\bar{E}M\}_{BC} = \begin{Bmatrix} 89.89 \\ -44.44 \\ 66.67 \\ 33.333 \\ 0 \\ 0 \end{Bmatrix}$$

The joint load vector is given by,

$$\{J_L\} = \begin{Bmatrix} -48.89 \\ 44.44 \\ 60 \\ -66.67 \\ 0 \\ -33.33 \\ -40 \\ 60 \\ 0 \\ 0 \\ 0 \\ 0 \end{Bmatrix}_{12 \times 1} \begin{matrix} ① \\ ② \\ ③ \\ ④ \\ ⑤ \\ ⑥ \\ ⑦ \\ ⑧ \\ ⑨ \\ ⑩ \\ ⑪ \\ ⑫ \end{matrix}$$

Planar Non-Orthogonal Structures

Thus, the joint load vector in unrestrained degrees-of-freedom is given by,

$$\{\bar{J}_{lu}\} = \begin{Bmatrix} -48.89 \\ 44.44 \\ 60 \\ -66.67 \\ 0 \\ -33.33 \end{Bmatrix}_{6\times 1} \begin{matrix} (1) \\ (2) \\ (3) \\ (4) \\ (5) \\ (6) \end{matrix}$$

5. *Calculation of the global stiffness matrix:*
   The stiffness matrix of every member, with respect to the global axes system, is obtained by the following equation:

$$\{\bar{K}\}_i = [T]_i^T [K]_i [T]_i$$

$$\bar{K}_{AB} = E \times 10^{-4} \begin{bmatrix} 23 & 11 & 0 & 0 & -9 & 9 \\ 11 & 23 & 0 & 0 & -9 & 9 \\ 0 & 0 & 338 & -338 & 0 & 0 \\ 0 & 0 & -338 & 338 & 0 & 0 \\ -9 & -9 & 0 & 0 & 4 & -4 \\ 9 & 9 & 0 & 0 & -4 & 4 \end{bmatrix} \begin{matrix} (7) \\ (1) \\ (9) \\ (4) \\ (8) \\ (3) \end{matrix}$$

columns: (7) (1) (9) (4) (8) (3)

$$\bar{K}_{BC} = E \times 10^{-4} \begin{bmatrix} 21 & 10 & 5 & -5 & 0 & 0 \\ 10 & 21 & 5 & -5 & 0 & 0 \\ 5 & 5 & 2 & -2 & 0 & 0 \\ -5 & -5 & -2 & 2 & 0 & 0 \\ 0 & 0 & 0 & 0 & 250 & -250 \\ 0 & 0 & 0 & 0 & -250 & 250 \end{bmatrix} \begin{matrix} (1) \\ (2) \\ (4) \\ (6) \\ (3) \\ (5) \end{matrix}$$

columns: (1) (2) (4) (6) (3) (5)

$$\bar{K}_{CD} = E \times 10^{-4} \begin{bmatrix} 18 & 9 & 3 & -3 & 4 & -4 \\ 9 & 18 & 3 & -3 & 4 & -4 \\ 3 & 3 & 174 & -174 & -129 & 129 \\ -3 & -3 & -174 & 174 & 129 & -129 \\ 4 & 4 & -129 & 129 & 99 & -99 \\ -4 & -4 & 129 & -129 & -99 & 99 \end{bmatrix} \begin{matrix} (2) \\ (10) \\ (6) \\ (12) \\ (5) \\ (11) \end{matrix}$$

columns: (2) (10) (6) (12) (5) (11)

$$\bar{K}_{TOTAL} = E \times 10^{-4} \begin{bmatrix} 44 & 10 & 9 & 5 & 0 & -5 & | & 11 & -9 & 0 & 0 & 0 & 0 \\ 10 & 39 & 0 & 5 & 4 & -2 & | & 0 & 0 & 0 & 9 & -4 & -3 \\ 9 & 0 & 254 & 0 & -250 & 0 & | & 9 & -4 & 0 & 0 & 0 & 0 \\ 5 & 5 & 0 & 339 & 0 & -2 & | & 0 & 0 & -338 & 0 & 0 & 0 \\ 0 & 4 & -250 & 0 & 349 & -129 & | & 0 & 0 & 0 & 4 & -99 & 129 \\ -5 & -2 & 0 & -2 & -129 & 175 & | & 0 & 0 & 0 & 3 & 129 & -174 \\ \hdashline 11 & 0 & 9 & 0 & 0 & 0 & | & 23 & -9 & 0 & 0 & 0 & 0 \\ -9 & 0 & -4 & 0 & 0 & 0 & | & -9 & 4 & 0 & 0 & 0 & 0 \\ 0 & 0 & 0 & -338 & 0 & 0 & | & 0 & 0 & 338 & 0 & 0 & 0 \\ 0 & 9 & 0 & 0 & 4 & 3 & | & 0 & 0 & 0 & 18 & -4 & -3 \\ 0 & -4 & 0 & 0 & -99 & 126 & | & 0 & 0 & 0 & -4 & 99 & -129 \\ 0 & -3 & 0 & 0 & 129 & -174 & | & 0 & 0 & 0 & -3 & -129 & 174 \end{bmatrix} \begin{matrix} ① \\ ② \\ ③ \\ ④ \\ ⑤ \\ ⑥ \\ ⑦ \\ ⑧ \\ ⑨ \\ ⑩ \\ ⑪ \\ ⑫ \end{matrix}$$

$$\left[\bar{K}_{uu}\right] = E \times 10^{-4} \begin{bmatrix} 44 & 10 & 9 & 5 & 0 & -5 \\ 10 & 39 & 0 & 5 & 4 & -2 \\ 9 & 0 & 254 & 0 & -250 & 0 \\ 5 & 5 & 0 & 339 & 0 & -2 \\ 0 & 4 & -250 & 0 & 349 & -120 \\ -5 & -2 & 0 & -2 & -129 & 175 \end{bmatrix} \begin{matrix} ① \\ ② \\ ③ \\ ④ \\ ⑤ \\ ⑥ \end{matrix}$$

$$\left[\bar{K}_{uu}\right]^{-1} = \frac{1}{E} \begin{bmatrix} 258.7 & -59.2 & -125.5 & -3.5 & -118.8 & -80.1 \\ -59.2 & 277.2 & -61.8 & -3.6 & -64.9 & -46.4 \\ -125.5 & -61.8 & 1262.3 & 7.5 & 1239.6 & 904.8 \\ -3.5 & -3.6 & 7.5 & 29.6 & 7.5 & 5.7 \\ -118.8 & -64.9 & 1239.6 & 7.5 & 1256.7 & 917.5 \\ -80.1 & -46.4 & 904.8 & 5.7 & 917.5 & 727.1 \end{bmatrix}$$

Now, $\{\bar{\delta}_u\} = \left[\bar{K}_{uu}\right]^{-1} \{\bar{J}_L\}_u$

$$\{\bar{\delta}_u\} = \frac{1}{E} \times 10^4 \begin{Bmatrix} -1.9906 \\ 1.3283 \\ 4.8465 \\ -0.1702 \\ 4.6213 \\ 3.1529 \end{Bmatrix}$$

$$\left[\bar{\delta}\right]_{AB} = \frac{1}{E} \times 10^4 \begin{Bmatrix} 0 \\ -1.9906 \\ 0 \\ -0.1702 \\ 0 \\ 4.8465 \end{Bmatrix}, \quad \left[\bar{\delta}\right]_{BC} = \frac{1}{E} \times 10^4 \begin{Bmatrix} -1.9906 \\ 1.3283 \\ -0.1702 \\ 3.1529 \\ 4.8465 \\ 4.6213 \end{Bmatrix}, \quad \left[\bar{\delta}\right]_{CD} = \frac{1}{E} \times 10^4 \begin{Bmatrix} 1.3283 \\ 0 \\ 3.1529 \\ 0 \\ 4.6213 \\ 0 \end{Bmatrix}$$

Planar Non-Orthogonal Structures

6. *Calculation of end moments and shear:*

$$\{\bar{M}\}_i = [\bar{K}]_i \{\bar{d}_i\} + \{\bar{FEM}\}_i$$

$$[\bar{M}]_{AB} = \begin{Bmatrix} \bar{M}_7 \\ \bar{M}_1 \\ \bar{V}_9 \\ \bar{V}_4 \\ \bar{H}_8 \\ \bar{H}_3 \end{Bmatrix} = \begin{Bmatrix} 58.745 \\ -43.9479 \\ 57.4513 \\ -57.4513 \\ -63.6993 \\ -56.3007 \end{Bmatrix}$$

$$[\bar{M}]_{BC} = \begin{Bmatrix} \bar{M}_1 \\ \bar{M}_2 \\ \bar{V}_4 \\ \bar{V}_6 \\ \bar{H}_3 \\ \bar{H}_5 \end{Bmatrix} = \begin{Bmatrix} 43.9479 \\ -54.8100 \\ 57.4513 \\ 42.5487 \\ 56.3007 \\ -56.3007 \end{Bmatrix}$$

$$[\bar{M}]_{CD} = \begin{Bmatrix} \bar{M}_2 \\ \bar{M}_{10} \\ \bar{V}_6 \\ \bar{V}_{12} \\ \bar{H}_5 \\ \bar{H}_{11} \end{Bmatrix} = \begin{Bmatrix} 54.8100 \\ 42.6958 \\ -42.5517 \\ 42.5517 \\ 56.3007 \\ -56.3007 \end{Bmatrix}$$

The member and final and moments are shown in Figures 2.22 and 2.23.

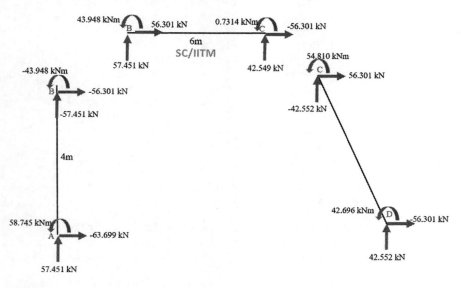

FIGURE 2.22 Member end moments and shear.

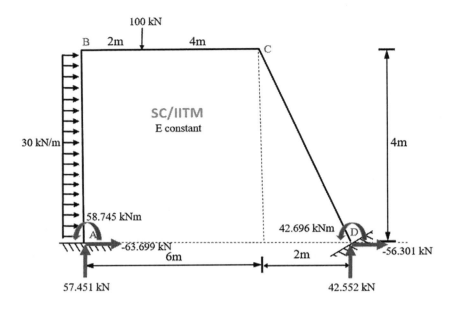

FIGURE 2.23 Final end moments and shear.

MATLAB program:

```
%% stiffness matrix method
% Input
clc;
clear;
n = 3; % number of members
I = [2.28E-3 3.125E-3 2.28E-3]; %Moment of inertis in m4
L = [4 6 5]; % length in m
A = [0.135 0.15 0.135]; % Area in m2
theta= [90 0 -53.123]; % angle in degrees
uu = 6; % Number of unrestrained degrees of freedom
ur = 6; % Number of restrained degrees of freedom
uul = [1 2 3 4 5 6]; % global labels of unrestrained dof
url = [7 8 9 10 11 12]; % global labels of restrained dof
l1 = [7 1 9 4 8 3]; % Global labels for member 1
l2 = [1 2 4 6 3 5]; % Global labels for member 2
l3 = [2 10 6 12 5 11]; % Global labels for member 3
l= [l1; l2; l3];
dof = uu + ur; % Degrees of freedom
Ktotal = zeros (dof);
Tt1 = zeros (6); % Transformation matrix for member 1
Tt2 = zeros (6); % Transformation matrix for member 2
Tt3 = zeros (6); % Transformation matrix for member 3
fem1= [40; -40; 60; 60; 0; 0]; % Local Fixed end moments of member 1
fem2= [88.89; -44.44; 66.67; 33.33; 0; 0]; % Local Fixed end moments of member 2
```

Planar Non-Orthogonal Structures

```
fem3= [0; 0; 0; 0; 0; 0]; % Local Fixed end moments of member 3

%% rotation coefficients for each member
rc1 = 4.*I./L;
rc2 = 2.*I./L;
rc3 = A./L;
cx = cosd(theta);
cy = sind(theta);

%% stiffness matrix 6 by 6
for i = 1:n
          Knew = zeros (dof);
          k1 = [rc1(i); rc2(i); (rc1(i)+rc2(i))/L(i);
(-(rc1(i)+rc2(i))/L(i)); 0; 0];
          k2 = [rc2(i); rc1(i); (rc1(i)+rc2(i))/L(i);
(-(rc1(i)+rc2(i))/L(i)); 0; 0;];
          k3 = [(rc1(i)+rc2(i))/L(i); (rc1(i)+rc2(i))/L(i);
(2*(rc1(i)+rc2(i))/(L(i)^2)); (-2*(rc1(i)+rc2(i))/(L(i)^2));
0; 0;];
          k4 = -k3;
          k5 = [0; 0; 0; 0; rc3(i); -rc3(i)];
          k6 = [0; 0; 0; 0; -rc3(i); rc3(i)];
          K = [k1 k2 k3 k4 k5 k6];
          fprintf ('Member Number =');
          disp (i);
          fprintf ('Local Stiffness matrix of member, [K] =
\n');
          disp (K);
          T1 = [1; 0; 0; 0; 0; 0];
          T2 = [0; 1; 0; 0; 0; 0];
          T3 = [0; 0; cx(i); 0; cy(i); 0];
          T4 = [0; 0; 0; cx(i); 0; cy(i)];
          T5 = [0; 0; -cy(i); 0; cx(i); 0];
          T6 = [0; 0; 0; -cy(i); 0; cx(i)];
          T = [T1 T2 T3 T4 T5 T6];
          fprintf ('Tranformation matrix of member, [T] = \n');
          disp (T);
          Ttr = T';
          fprintf ('Tranformation matrix Transpose, [T] = \n');
          disp (Ttr);
          Kg = Ttr*K*T;
          fprintf ('Global Matrix, [K global] = \n');
          disp (Kg);
          for p = 1:6
              for q = 1:6
                  Knew((l(i,p)),(l(i,q))) =Kg(p,q);
              end
          end
          Ktotal = Ktotal + Knew;
          if i == 1
              Tt1= T;
```

```
                Kg1=Kg;
                fembar1= Tt1'*fem1;
            elseif i == 2
                Tt2 = T;
                Kg2 = Kg;
                fembar2= Tt2'*fem2;
            else
                Tt3 = T;
                Kg3=Kg;
                fembar3= Tt3'*fem3;
            end
end
fprintf ('Stiffness Matrix of complete structure, [Ktotal] = \n');
disp (Ktotal);
Kunr = zeros(6);
for x=1:uu
        for y=1:uu
            Kunr(x,y)= Ktotal(x,y);
        end
end
fprintf ('Unrestrained Stiffness sub-matrix, [Kuu] = \n');
disp (Kunr);
KuuInv= inv(Kunr);
fprintf ('Inverse of Unrestrained Stiffness sub-matrix,
[KuuInverse] = \n');
disp (KuuInv);

%% Creation of joint load vector
jl= [-48.89; 44.44; 60; -66.67; 0; -33.333; -40; 60; 0; 0; 0;
0]; % values given in kN or kNm
jlu = jl(1:uu,1); % load vector in unrestrained dof
delu = KuuInv*jlu;
fprintf ('Joint Load vector, [Jl] = \n');
disp (jl');
fprintf ('Unrestrained displacements, [DelU] = \n');
disp (delu');
delr = zeros (ur,1);
del = zeros (dof,1);
del = [delu; delr];
deli= zeros (6,1);
for i = 1:n
    for p = 1:6
        deli(p,1) = del((l(i,p)),1) ;
    end
    if i == 1
            delbar1 = deli;
            mbar1= (Kg1 * delbar1)+fembar1;
            fprintf ('Member Number =');
            disp (i);
            fprintf ('Global displacement matrix [DeltaBar] = \n');
            disp (delbar1');
```

Planar Non-Orthogonal Structures

```
                fprintf ('Global End moment matrix [MBar] = \n');
                disp (mbar1');
            elseif i == 2
                delbar2 = deli;
                mbar2= (Kg2 * delbar2)+fembar2;
                fprintf ('Member Number =');
                disp (i);
                fprintf ('Global displacement matrix [DeltaBar] = \n');
                disp (delbar2');
                fprintf ('Global End moment matrix [MBar] = \n');
                disp (mbar2');
            else
                delbar3 = deli;
                mbar3= (Kg3 * delbar3)+fembar3;
                fprintf ('Member Number =');
                disp (i);
                fprintf ('Global displacement matrix [DeltaBar] = \n');
                disp (delbar3');
                fprintf ('Global End moment matrix [MBar] = \n');
                disp (mbar3');
        end
end

%% check
mbar = [mbar1'; mbar2'; mbar3'];
jf = zeros(dof,1);
for a=1:n
    for b=1:6 % size of k matrix
        d = l(a,b);
        jfnew = zeros(dof,1);
        jfnew(d,1)=mbar(a,b);
        jf=jf+jfnew;
    end
end
fprintf ('Joint forces = \n');
disp (jf');
```

MATLAB output:

```
Member Number = 1
Local Stiffness matrix of member, [K] =

    0.0023    0.0011    0.0009   -0.0009         0         0
    0.0011    0.0023    0.0009   -0.0009         0         0
    0.0009    0.0009    0.0004   -0.0004         0         0
   -0.0009   -0.0009   -0.0004    0.0004         0         0
         0         0         0         0    0.0338   -0.0338
         0         0         0         0   -0.0338    0.0338
```

Tranformation matrix of member, [T] =

```
1  0  0  0   0   0
0  1  0  0   0   0
0  0  0  0  -1   0
0  0  0  0   0  -1
0  0  1  0   0   0
0  0  0  1   0   0
```

Tranformation matrix Transpose, [T] =

```
1  0   0   0  0  0
0  1   0   0  0  0
0  0   0   0  1  0
0  0   0   0  0  1
0  0  -1   0  0  0
0  0   0  -1  0  0
```

Global Matrix, [K global] =

```
 0.0023   0.0011    0         0       -0.0009   0.0009
 0.0011   0.0023    0         0       -0.0009   0.0009
 0        0         0.0338   -0.0338   0        0
 0        0        -0.0338    0.0338   0        0
-0.0009  -0.0009    0         0        0.0004  -0.0004
 0.0009   0.0009    0         0       -0.0004   0.0004
```

Member Number = 2
Local Stiffness matrix of member, [K] =

```
 0.0021   0.0010   0.0005  -0.0005   0        0
 0.0010   0.0021   0.0005  -0.0005   0        0
 0.0005   0.0005   0.0002  -0.0002   0        0
-0.0005  -0.0005  -0.0002   0.0002   0        0
 0        0        0        0        0.0250  -0.0250
 0        0        0        0       -0.0250   0.0250
```

Tranformation matrix of member, [T] =

```
1  0  0  0  0  0
0  1  0  0  0  0
0  0  1  0  0  0
0  0  0  1  0  0
0  0  0  0  1  0
0  0  0  0  0  1
```

Planar Non-Orthogonal Structures

Tranformation matrix Transpose, [T] =

```
1 0 0 0 0 0
0 1 0 0 0 0
0 0 1 0 0 0
0 0 0 1 0 0
0 0 0 0 1 0
0 0 0 0 0 1
```

Global Matrix, [K global] =

```
 0.0021   0.0010   0.0005  -0.0005      0        0
 0.0010   0.0021   0.0005  -0.0005      0        0
 0.0005   0.0005   0.0002  -0.0002      0        0
-0.0005  -0.0005  -0.0002   0.0002      0        0
    0        0        0        0     0.0250  -0.0250
    0        0        0        0    -0.0250   0.0250
```

Member Number =   3
Local Stiffness matrix of member, [K] =

```
 0.0018   0.0009   0.0005  -0.0005      0        0
 0.0009   0.0018   0.0005  -0.0005      0        0
 0.0005   0.0005   0.0002  -0.0002      0        0
-0.0005  -0.0005  -0.0002   0.0002      0        0
    0        0        0        0     0.0270  -0.0270
    0        0        0        0    -0.0270   0.0270
```

Tranformation matrix of member, [T] =

```
1.0000      0        0        0        0        0
    0   1.0000      0        0        0        0
    0        0    0.6001      0     0.7999      0
    0        0        0     0.6001      0     0.7999
    0        0   -0.7999      0     0.6001      0
    0        0        0    -0.7999      0     0.6001
```

Tranformation matrix Transpose, [T] =

```
1.0000      0        0        0        0        0
    0   1.0000      0        0        0        0
    0        0    0.6001      0    -0.7999      0
    0        0        0     0.6001      0    -0.7999
    0        0    0.7999      0     0.6001      0
    0        0        0     0.7999      0     0.6001
```

Global Matrix, [K global] =

```
   0.0018    0.0009    0.0003   -0.0003    0.0004   -0.0004
   0.0009    0.0018    0.0003   -0.0003    0.0004   -0.0004
   0.0003    0.0003    0.0174   -0.0174   -0.0129    0.0129
  -0.0003   -0.0003   -0.0174    0.0174    0.0129   -0.0129
   0.0004    0.0004   -0.0129    0.0129    0.0099   -0.0099
  -0.0004   -0.0004    0.0129   -0.0129   -0.0099    0.0099
```

Stiffness Matrix of complete structure, [Ktotal] =
Columns 1 through 10

```
   0.0044    0.0010    0.0009    0.0005         0   -0.0005    0.0011   -0.0009         0         0
   0.0010    0.0039         0    0.0005    0.0004   -0.0002         0         0         0    0.0009
   0.0009         0    0.0254         0   -0.0250         0    0.0009   -0.0004         0         0
   0.0005    0.0005         0    0.0339         0   -0.0002         0         0   -0.0338         0
        0    0.0004   -0.0250         0    0.0349   -0.0129         0         0         0    0.0004
  -0.0005   -0.0002         0   -0.0002   -0.0129    0.0175         0         0         0    0.0003
   0.0011         0    0.0009         0         0         0    0.0023   -0.0009         0         0
  -0.0009         0   -0.0004         0         0         0   -0.0009    0.0004         0         0
        0         0         0   -0.0338         0         0         0         0    0.0338         0
        0    0.0009         0         0    0.0004    0.0003         0         0         0    0.0018
        0   -0.0004         0         0   -0.0099    0.0129         0         0         0   -0.0004
        0   -0.0003         0         0    0.0129   -0.0174         0         0         0   -0.0003
```

Columns 11 through 12

```
        0         0
  -0.0004   -0.0003
        0         0
        0         0
  -0.0099    0.0129
   0.0129   -0.0174
        0         0
        0         0
        0         0
  -0.0004   -0.0003
   0.0099   -0.0129
  -0.0129    0.0174
```

Unrestrained Stiffness sub-matrix, [Kuu] =

```
   0.0044    0.0010    0.0009    0.0005         0   -0.0005
   0.0010    0.0039         0    0.0005    0.0004   -0.0002
   0.0009         0    0.0254         0   -0.0250         0
   0.0005    0.0005         0    0.0339         0   -0.0002
        0    0.0004   -0.0250         0    0.0349   -0.0129
  -0.0005   -0.0002         0   -0.0002   -0.0129    0.0175
```

Planar Non-Orthogonal Structures

Inverse of Unrestrained Stiffness sub-matrix, [KuuInverse] =

```
1.0e+03 *
   0.2587   -0.0592   -0.1255   -0.0035   -0.1188   -0.0801
  -0.0592    0.2772   -0.0618   -0.0036   -0.0649   -0.0464
  -0.1255   -0.0618    1.2623    0.0075    1.2396    0.9048
  -0.0035   -0.0036    0.0075    0.0296    0.0075    0.0057
  -0.1188   -0.0649    1.2396    0.0075    1.2567    0.9175
  -0.0801   -0.0464    0.9048    0.0057    0.9175    0.7271
```

Joint Load vector, [Jl] =

```
 -48.8900   44.4400   60.0000  -66.6700        0  -33.3330
 -40.0000   60.0000        0         0         0         0
```

Unrestrained displacements, [DelU] =

```
1.0e+04 *
 -1.9906    1.3283    4.8465   -0.1702    4.6213    3.1529
```

Member Number =   1
Global displacement matrix [DeltaBar] =

```
1.0e+04 *
       0   -1.9906        0   -0.1702        0    4.8465
```

Global End moment matrix [MBar] =

```
  58.7450  -43.9479   57.4513  -57.4513  -63.6993  -56.3007
```

Member Number =   2
Global displacement matrix [DeltaBar] =

```
1.0e+04 *
 -1.9906    1.3283   -0.1702    3.1529    4.8465    4.6213
```

Global End moment matrix [MBar] =

```
  43.9479  -54.8100   57.4513   42.5487   56.3007  -56.3007
```

Member Number =   3
Global displacement matrix [DeltaBar] =

```
1.0e+04 *
  1.3283         0    3.1529         0    4.6213         0
```

Global End moment matrix [MBar] =

```
  54.8100   42.6958  -42.5517   42.5517   56.3007  -56.3007
```

Joint forces =

```
   0.0000    0.0000   -0.0000   -0.0000    0.0000   -0.0030
  58.7450  -63.6993   57.4513   42.6958  -56.3007   42.5517
```

3 Planar Truss Structures

3.1 PLANAR TRUSS SYSTEM

In a truss system, joints are assumed to be pinned, which means that there is no moment transfer. Therefore, they can only resist the axial force and axial deformation. Thus, at every node or joint in a truss system, there are only two possible independent components of joint translation with respect to the reference axes system. Similarly, in the local axes system, each joint can have only two joint translation. Thus, there is no rotations at the ends.

3.1.1 Transformation Matrix

The following Figure 3.1 explains the transformation between the local and global axes. Consider a truss member inclined to an arbitrary angle of θ. This angle should always be measured with respect to the global axes (X-Y). There will be independent translations happening along X and Y directions.

We also know that,

$C_x = \cos\theta$, $C_y = \sin\theta$.

In order to convert the local responses with respect to the reference axes responses, the transformation matrix can be written as follows:

$$\begin{Bmatrix} p_r \\ p_s \\ p_t \\ p_h \end{Bmatrix} = \begin{bmatrix} C_x & 0 & -C_y & 0 \\ 0 & C_x & 0 & -C_y \\ C_y & 0 & C_x & 0 \\ 0 & C_y & 0 & C_x \end{bmatrix} \begin{Bmatrix} \bar{p}_r \\ \bar{p}_s \\ \bar{p}_t \\ \bar{p}_h \end{Bmatrix} \tag{3.1}$$

$$\{p\}_i = [T_T]_i \{\bar{p}\}_i$$

Hence,

$$\{\delta_T\}_i = [T_T]_i \{\bar{\delta}_T\}_i$$

$$\{\bar{\delta}_T\}_i = [T_T]_i^T \{\delta_T\}_i \tag{3.2}$$

where,

$$\{\delta_T\}_i = \begin{Bmatrix} \delta_r \\ \delta_s \\ \delta_t \\ \delta_h \end{Bmatrix}, \quad \{\bar{\delta}_T\}_i = \begin{Bmatrix} \bar{\delta}_r \\ \bar{\delta}_s \\ \bar{\delta}_t \\ \bar{\delta}_h \end{Bmatrix}$$

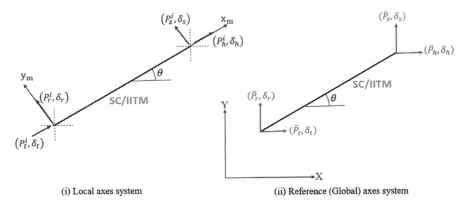

FIGURE 3.1 Transformation between local and global axes.

3.1.2 Stiffness Matrix

Now, the standard stiffness matrix of the truss element without end rotations is simply given as follows:

$$[K_T]_i = \begin{bmatrix} 0 & 0 & 0 & 0 \\ 0 & 0 & 0 & 0 \\ 0 & 0 & \dfrac{AE}{l} & -\dfrac{AE}{l} \\ 0 & 0 & -\dfrac{AE}{l} & \dfrac{AE}{l} \end{bmatrix}$$

Now, the global stiffness matrix of the truss member can be obtained using the following equation:

$$[\bar{K}_T]_i = [T_T]_i^T [K_T]_i [T_T]_i \qquad (3.3)$$

The responses of the truss member in the reference axes system is given by,

$$[\bar{P}_T]_i = [\bar{K}_T]_i \{\bar{\delta}_T\}_i + \{\bar{F}_p\}_i \qquad (3.4)$$

where,

$$[\bar{P}_T]_i = \begin{Bmatrix} \bar{p}_r \\ \bar{p}_s \\ \bar{p}_s \\ \bar{p}_h \end{Bmatrix}, \quad \{\bar{F}_p\}_i = \begin{Bmatrix} \bar{F}_{ps} \\ \bar{F}_{pr} \\ \bar{F}_{pt} \\ \bar{F}_{ph} \end{Bmatrix}_i$$

Hence, the planar truss problem is much simpler than the orthogonal and non-orthogonal structures problem.

Planar Truss Structures

Example problems with computer program

EXAMPLE 3.1:

Analyze the planar truss system shown in Figure 3.2 using the stiffness method. E is constant for all the members.

SOLUTION:

1. *Marking restrained and unrestrained degrees-of-freedom:*
   Unrestrained degrees-of-freedom = 4 (δ_1, δ_2, δ_3, δ_4)
   Restrained degrees-of-freedom = 4 (δ_5, δ_6, δ_7, δ_8)
   The global axes system and the local axes for every member should be marked to find θ.
   The unrestrained and restrained degrees-of-freedom are marked. The local and global axes system are also shown in Figure 3.3.

Member	Ends j	Ends k	Length (m)	θ (degrees)	C_x	C_y	Global Labels
AB	A	B	4	90	0	1	(6,2,5,1)
BC	B	C	4	0	1	0	(2,4,1,3)
CD	C	D	4	90	0	1	(8,4,7,3)
BD	B	D	5.646	−45	0.707	−0.707	(2,8,1,7)
AC	A	C	5.646	45	0.707	−0.707	(6,4,5,3)

2. *Calculation of local stiffness matrix:*
   The local stiffness matrix for the truss element is given by,

$$[K_T]_i = \begin{bmatrix} 0 & 0 & 0 & 0 \\ 0 & 0 & 0 & 0 \\ 0 & 0 & \frac{AE}{l} & -\frac{AE}{l} \\ 0 & 0 & -\frac{AE}{l} & \frac{AE}{l} \end{bmatrix}$$

$$K_{AB} = E \times 10^{-3} \begin{matrix} & \text{⑥} & \text{②} & \text{⑤} & \text{①} & \\ & \begin{bmatrix} 0 & 0 & 0 & 0 \\ 0 & 0 & 0 & 0 \\ 0 & 0 & 1.25 & -1.25 \\ 0 & 0 & -1.25 & 1.25 \end{bmatrix} & \begin{matrix} \text{⑥} \\ \text{②} \\ \text{⑤} \\ \text{①} \end{matrix} \end{matrix}$$

$$K_{BC} = E \times 10^{-3} \begin{matrix} & \text{②} & \text{④} & \text{①} & \text{③} & \\ & \begin{bmatrix} 0 & 0 & 0 & 0 \\ 0 & 0 & 0 & 0 \\ 0 & 0 & 1 & -1 \\ 0 & 0 & -1 & 1 \end{bmatrix} & \begin{matrix} \text{②} \\ \text{④} \\ \text{①} \\ \text{③} \end{matrix} \end{matrix}$$

126   Advanced Structural Analysis with MATLAB®

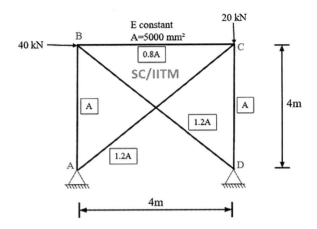

FIGURE 3.2 Truss example.

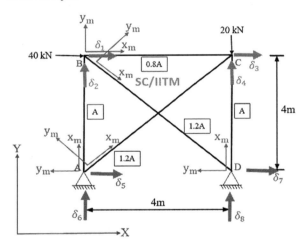

FIGURE 3.3 Marking degrees-of-freedom.

$$K_{CD} = E \times 10^{-3} \begin{bmatrix} 0 & 0 & 0 & 0 \\ 0 & 0 & 0 & 0 \\ 0 & 0 & 1.25 & -1.25 \\ 0 & 0 & -1.25 & 1.25 \end{bmatrix} \begin{matrix} \text{⑧} \\ \text{④} \\ \text{⑦} \\ \text{③} \end{matrix}$$

$$\begin{matrix} \text{⑧} & \text{④} & \text{⑦} & \text{③} \end{matrix}$$

$$K_{BD} = E \times 10^{-3} \begin{bmatrix} 0 & 0 & 0 & 0 \\ 0 & 0 & 0 & 0 \\ 0 & 0 & 1.061 & -1.061 \\ 0 & 0 & -1.061 & 1.061 \end{bmatrix} \begin{matrix} \text{②} \\ \text{⑧} \\ \text{①} \\ \text{⑦} \end{matrix}$$

$$\begin{matrix} \text{②} & \text{⑧} & \text{①} & \text{⑦} \end{matrix}$$

Planar Truss Structures

$$K_{AC} = E \times 10^{-3} \begin{bmatrix} 0 & 0 & 0 & 0 \\ 0 & 0 & 0 & 0 \\ 0 & 0 & 1.061 & -1.061 \\ 0 & 0 & -1.061 & 1.061 \end{bmatrix} \begin{matrix} 6 \\ 4 \\ 5 \\ 3 \end{matrix}$$

(column labels: 6, 4, 5, 3)

3. *Calculation of transformation matrix:*

$$[T]_{AB} = \begin{bmatrix} 0 & 0 & -1 & 0 \\ 0 & 0 & 0 & -1 \\ 1 & 0 & 0 & 0 \\ 0 & 1 & 0 & 0 \end{bmatrix}, \quad [T]_{BC} = \begin{bmatrix} 1 & 0 & 0 & 0 \\ 0 & 1 & 0 & 0 \\ 0 & 0 & 1 & 0 \\ 0 & 0 & 0 & 1 \end{bmatrix}, \quad [T]_{CD} = \begin{bmatrix} 0 & 0 & -1 & 0 \\ 0 & 0 & 0 & -1 \\ 1 & 0 & 0 & 0 \\ 0 & 1 & 0 & 0 \end{bmatrix}$$

$$[T]_{BD} = \begin{bmatrix} 0.7071 & 0 & 0.7071 & 0 \\ 0 & 0.7071 & 0 & 0.7071 \\ -0.7071 & 0 & 0.7071 & 0 \\ 0 & -0.7071 & 0 & 0.7071 \end{bmatrix},$$

$$[T]_{AC} = \begin{bmatrix} 0.7071 & 0 & -0.7071 & 0 \\ 0 & 0.7071 & 0 & -0.7071 \\ 0.7071 & 0 & 0.7071 & 0 \\ 0 & 0.7071 & 0 & 0.7071 \end{bmatrix}$$

4. *Calculation of the global stiffness matrix:*
The global stiffness matrix of all the members are calculated by using the following relationship:

$$\left[\bar{K}_T \right]_i = [T_T]_i^T [K_T]_i [T_T]_i$$

$$\bar{K}_{AB} = E \begin{bmatrix} 0.0013 & -0.0013 & 0 & 0 \\ -0.0013 & 0.0013 & 0 & 0 \\ 0 & 0 & 0 & 0 \\ 0 & 0 & 0 & 0 \end{bmatrix} \begin{matrix} 6 \\ 2 \\ 5 \\ 1 \end{matrix}$$

(column labels: 6, 2, 5, 1)

$$\bar{K}_{BC} = E \begin{bmatrix} 0 & 0 & 0 & 0 \\ 0 & 0 & 0 & 0 \\ 0 & 0 & 1 & -1 \\ 0 & 0 & -1 & 1 \end{bmatrix} \begin{matrix} 2 \\ 4 \\ 1 \\ 3 \end{matrix}$$

(column labels: 2, 4, 1, 3)

$$\bar{K}_{CD} = E \begin{bmatrix} 0.0013 & -0.0013 & 0 & 0 \\ -0.0013 & 0.0013 & 0 & 0 \\ 0 & 0 & 0 & 0 \\ 0 & 0 & 0 & 0 \end{bmatrix} \begin{matrix} 8 \\ 4 \\ 7 \\ 3 \end{matrix}$$

(column labels: 8, 4, 7, 3)

$$\bar{K}_{BD} = E \times 10^{-3} \begin{array}{c} \\ \begin{array}{cccc} \textcircled{2} & \textcircled{8} & \textcircled{1} & \textcircled{7} \end{array} \\ \begin{bmatrix} 0.5304 & -0.5304 & -0.5304 & 0.5304 \\ -0.5304 & 0.5304 & 0.5304 & -0.5304 \\ -0.5304 & 0.5304 & 0.5304 & -0.5304 \\ 0.5304 & -0.5304 & -0.5304 & 0.5304 \end{bmatrix} \begin{array}{c} \textcircled{2} \\ \textcircled{8} \\ \textcircled{1} \\ \textcircled{7} \end{array} \end{array}$$

$$\bar{K}_{AC} = E \times 10^{-3} \begin{array}{c} \\ \begin{array}{cccc} \textcircled{6} & \textcircled{4} & \textcircled{5} & \textcircled{3} \end{array} \\ \begin{bmatrix} 0.5304 & -0.5304 & 0.5304 & -0.5304 \\ -0.5304 & 0.5304 & -0.5304 & 0.5304 \\ 0.5304 & -0.5304 & 0.5304 & -0.5304 \\ -0.5304 & 0.5304 & -0.5304 & 0.5304 \end{bmatrix} \begin{array}{c} \textcircled{6} \\ \textcircled{4} \\ \textcircled{5} \\ \textcircled{3} \end{array} \end{array}$$

The total stiffness matrix is obtained by assembling the global stiffness matrices of all members with respect to the global labels. From the total stiffness matrix, the stiffness matrix for the unrestrained degrees-of-freedom can be partitioned.

$$[K_{uu}] = E \begin{array}{c} \begin{array}{cccc} \textcircled{1} & \textcircled{2} & \textcircled{3} & \textcircled{4} \end{array} \\ \begin{bmatrix} 0.0015 & -0.005 & -0.0010 & 0 \\ -0.005 & 0.0018 & 0 & 0 \\ -0.0010 & 0 & 0.0015 & 0.0005 \\ 0 & 0 & 0.0005 & 0.0018 \end{bmatrix} \begin{array}{c} \textcircled{1} \\ \textcircled{2} \\ \textcircled{3} \\ \textcircled{4} \end{array} \end{array}$$

$$[K_{uu}]^{-1} = \frac{1}{E} \times 10^{-3} \begin{bmatrix} 1.5534 & 0.4628 & 1.1319 & -0.3372 \\ 0.4628 & 0.6995 & 0.3372 & -0.1005 \\ 1.1319 & 0.3372 & 1.5534 & -0.4628 \\ -0.3372 & -0.1005 & -0.4628 & 0.6995 \end{bmatrix}$$

5. *Estimation of joint load vectors:*
The joint load vector can be written as follows:

$$[\bar{J}_L] = \begin{Bmatrix} +40 \\ 0 \\ 0 \\ -20 \\ 0 \\ 0 \\ 0 \\ 0 \end{Bmatrix} \begin{array}{c} \textcircled{1} \\ \textcircled{2} \\ \textcircled{3} \\ \textcircled{4} \\ \textcircled{5} \\ \textcircled{6} \\ \textcircled{7} \\ \textcircled{8} \end{array}$$

The joint load vector in unrestrained degrees-of-freedom can be partitioned from the previous matrix as follows:

Planar Truss Structures

$$[\bar{J}_{Lu}] = \begin{Bmatrix} +40 \\ 0 \\ 0 \\ -20 \end{Bmatrix} \begin{matrix} ① \\ ② \\ ③ \\ ④ \end{matrix}$$

$$\{\bar{\delta}\} = [\bar{K}_{uu}]^{-1}\{\bar{J}_{Lu}\}$$

$$\{\bar{\delta}\}_u = \frac{1\times10^{-4}}{E}\begin{Bmatrix} 6.881 \\ 2.0521 \\ 5.4532 \\ -2.7479 \end{Bmatrix}$$

6. Calculation of member forces:

$$\bar{M}_{AB} = \begin{Bmatrix} \bar{V}_6 \\ \bar{V}_2 \\ \bar{H}_5 \\ \bar{H}_1 \end{Bmatrix} = \begin{Bmatrix} -25.6509 \\ 25.6509 \\ 0 \\ 0 \end{Bmatrix}, \bar{M}_{BC} = \begin{Bmatrix} \bar{V}_2 \\ \bar{V}_4 \\ \bar{H}_1 \\ \bar{H}_5 \end{Bmatrix} = \begin{Bmatrix} 0 \\ 0 \\ 14.3491 \\ -14.3491 \end{Bmatrix}, \bar{M}_{CD} = \begin{Bmatrix} \bar{V}_8 \\ \bar{V}_4 \\ \bar{H}_7 \\ \bar{H}_3 \end{Bmatrix} = \begin{Bmatrix} 34.3491 \\ -34.3491 \\ 0 \\ 0 \end{Bmatrix}$$

$$\bar{M}_{BD} = \begin{Bmatrix} \bar{V}_2 \\ \bar{V}_8 \\ \bar{H}_1 \\ \bar{H}_7 \end{Bmatrix} = \begin{Bmatrix} -25.6509 \\ 25.6509 \\ 26.6509 \\ -25.6509 \end{Bmatrix}, \bar{M}_{AC} = \begin{Bmatrix} \bar{V}_6 \\ \bar{V}_4 \\ \bar{H}_5 \\ \bar{H}_3 \end{Bmatrix} = \begin{Bmatrix} -14.3491 \\ 14.3491 \\ -14.3491 \\ 14.3491 \end{Bmatrix}$$

The member and find end forces are shown in Figures 3.4 and 3.5.

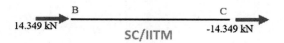

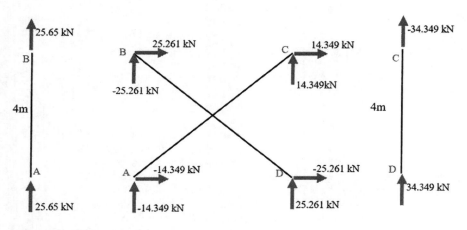

FIGURE 3.4 Member end forces.

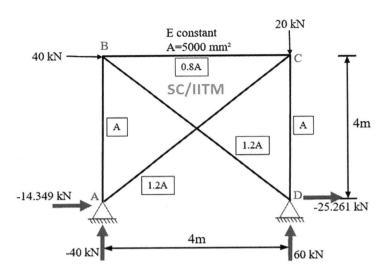

FIGURE 3.5 Final end forces.

MATLAB® program:

```
%% stiffness matrix method for planar truss
% Input
clc;
clear;
n = 5; % number of members
L = [4 4 4 5.646 5.646]; % length in m
A = [5E-3 4E-3 5E-3 6E-3 6E-3]; % Area in m2
theta= [90 0 90 -45 45]; % angle in degrees
uu = 4; % Number of unrestrained degrees-of-freedom
ur = 4; % Number of restrained degrees-of-freedom
uul = [1 2 3 4]; % global labels of unrestrained dof
url = [5 6 7 8]; % global labels of restrained dof
l1 = [6 2 5 1]; % Global labels for member 1
l2 = [2 4 1 3]; % Global labels for member 2
l3 = [8 4 7 3]; % Global labels for member 3
l4 = [2 8 1 7]; % Global labels for member 4
l5 = [6 4 5 3]; % Global labels for member 5
l= [l1; l2; l3; l4; l5];
dof = uu + ur; % Degrees-of-freedom
Ktotal = zeros (dof);
Tt1 = zeros (4); % Transformation matrix for member 1
Tt2 = zeros (4); % Transformation matrix for member 2
Tt3 = zeros (4); % Transformation matrix for member 3
Tt4 = zeros (4); % Transformation matrix for member 4
Tt5 = zeros (4); % Transformation matrix for member 5
fem1= [0; 0; 0; 0]; % Local Fixed end forces of member 1
fem2= [0; 0; 0; 0]; % Local Fixed end forces of member 2
fem3= [0; 0; 0; 0]; % Local Fixed end forces of member 3
```

Planar Truss Structures

```
fem4= [0; 0; 0; 0]; % Local Fixed end forces of member 4
fem5= [0; 0; 0; 0]; % Local Fixed end forces of member 5

%% rotation coefficients for each member
rc = A./L;
cx = cosd(theta);
cy = sind(theta);

%% stiffness matrix 4 by 4
for i = 1:n
      Knew = zeros (dof);
      k1 = [0; 0; 0; 0];
      k2 = [0; 0; 0; 0];
      k3 = [0; 0; rc(i); -rc(i)];
      k4 = -k3;
      K = [k1 k2 k3 k4];
      fprintf ('Member Number =');
      disp (i);
      fprintf ('Local Stiffness matrix of member, [K] = \n');
      disp (K);
      T1 = [cx(i); 0; cy(i); 0];
      T2 = [0; cx(i); 0; cy(i)];
      T3 = [-cy(i); 0; cx(i); 0];
      T4 = [0; -cy(i); 0; cx(i)];
      T = [T1 T2 T3 T4];
      fprintf ('Tranformation matrix of member, [T] = \n');
      disp (T);
      Ttr = T';
      fprintf ('Tranformation matrix Transpose, [T] = \n');
      disp (Ttr);
      Kg = Ttr*K*T;
      fprintf ('Global Matrix, [K global] = \n');
      disp (Kg);
      for p = 1:4
         for q = 1:4
            Knew((l(i,p)),(l(i,q))) =Kg(p,q);
         end
      end
      Ktotal = Ktotal + Knew;
      if i == 1
         Tt1= T;
         Kg1=Kg;
         fembar1= Tt1'*fem1;
      elseif i == 2
         Tt2 = T;
         Kg2 = Kg;
         fembar2= Tt2'*fem2;
      elseif i == 3
         Tt3 = T;
         Kg3 = Kg;
         fembar3= Tt3'*fem3;
```

```
            elseif i == 4
               Tt4 = T;
               Kg4 = Kg;
               fembar4= Tt4'*fem4;
            else
               Tt5 = T;
               Kg5=Kg;
               fembar5= Tt5'*fem5;
            end
end
fprintf ('Stiffness Matrix of complete structure, [Ktotal] = \n');
disp (Ktotal);
Kunr = zeros(uu);
for x=1:uu
   for y=1:uu
      Kunr(x,y)= Ktotal(x,y);
   end
end
fprintf ('Unrestrained Stiffness sub-matrix, [Kuu] = \n');
disp (Kunr);
KuuInv= inv(Kunr);
fprintf ('Inverse of Unrestrained Stiffness sub-matrix,
[KuuInverse] = \n');
disp (KuuInv);

%% Creation of joint load vector
jl= [40; 0; 0; -20; 0; 0; 0; 0]; % values given in kN
jlu = [40; 0; 0; -20]; % load vector in unrestrained dof
delu = KuuInv*jlu;
fprintf ('Joint Load vector, [Jl] = \n');
disp (jl');
fprintf ('Unrestrained displacements, [DelU] = \n');
disp (delu');
delr = zeros(ur,1);
del = zeros (dof,1);
del = [delu; delr];
deli= zeros (4,1);
for i = 1:n
   for p = 1:4
      deli(p,1) = del((l(i,p)),1) ;
   end
   if i == 1
         delbar1 = deli;
         mbar1= (Kg1 * delbar1)+fembar1;
         fprintf ('Member Number =');
         disp (i);
         fprintf ('Global displacement matrix [DeltaBar] = \n');
         disp (delbar1');
         fprintf ('Global End moment matrix [MBar] = \n');
         disp (mbar1');
      elseif i == 2
```

```
            delbar2 = deli;
            mbar2= (Kg2 * delbar2)+fembar2;
            fprintf ('Member Number =');
            disp (i);
            fprintf ('Global displacement matrix [DeltaBar] = \n');
            disp (delbar2');
            fprintf ('Global End moment matrix [MBar] = \n');
            disp (mbar2');
        elseif i ==3
            delbar3 = deli;
            mbar3= (Kg3 * delbar3)+fembar3;
            fprintf ('Member Number =');
            disp (i);
            fprintf ('Global displacement matrix [DeltaBar] = \n');
            disp (delbar3');
            fprintf ('Global End moment matrix [MBar] = \n');
            disp (mbar3');
        elseif i == 4
            delbar4 = deli;
            mbar4= (Kg4 * delbar4)+fembar4;
            fprintf ('Member Number =');
            disp (i);
            fprintf ('Global displacement matrix [DeltaBar] = \n');
            disp (delbar4');
            fprintf ('Global End moment matrix [MBar] = \n');
            disp (mbar4');
        else
            delbar5 = deli;
            mbar5= (Kg5 * delbar5)+fembar5;
            fprintf ('Member Number =');
            disp (i);
            fprintf ('Global displacement matrix [DeltaBar] = \n');
            disp (delbar5');
            fprintf ('Global End moment matrix [MBar] = \n');
            disp (mbar5');
     end
end

%% check
mbar = [mbar1'; mbar2'; mbar3'; mbar4'; mbar5'];
jf = zeros(dof,1);
for a=1:n
    for b=1:4 % size of k matrix
        d = l(a,b);
        jfnew = zeros(dof,1);
        jfnew(d,1)=mbar(a,b);
        jf=jf+jfnew;
    end
end
fprintf ('Joint forces = \n');
disp (jf');
```

MATLAB® output:

```
Member Number = 1
Local Stiffness matrix of member, [K] =

        0    0       0         0
        0    0       0         0
        0    0    0.0013    -0.0013
        0    0   -0.0013     0.0013

Transformation matrix of member, [T] =

        0    0   -1    0
        0    0    0   -1
        1    0    0    0
        0    1    0    0

Transformation matrix Transpose, [T] =

        0    0    1    0
        0    0    0    1
       -1    0    0    0
        0   -1    0    0

Global Matrix, [K global] =

     0.0013   -0.0013    0    0
    -0.0013    0.0013    0    0
          0         0    0    0
          0         0    0    0

Member Number = 2
Local Stiffness matrix of member, [K] =

    1.0e-03 *

        0       0       0         0
        0       0       0         0
        0       0    1.0000    -1.0000
        0       0   -1.0000     1.0000

Transformation matrix of member, [T] =

        1    0    0    0
        0    1    0    0
        0    0    1    0
        0    0    0    1

Transformation matrix Transpose, [T] =

        1    0    0    0
        0    1    0    0
        0    0    1    0
        0    0    0    1
```

Planar Truss Structures

```
Global Matrix, [K global] =

    1.0e-03 *
            0       0        0        0
            0       0        0        0
            0       0   1.0000  -1.0000
            0       0  -1.0000   1.0000

Member Number = 3
Local Stiffness matrix of member, [K] =

      0   0        0        0
      0   0        0        0
      0   0   0.0013  -0.0013
      0   0  -0.0013   0.0013

Transformation matrix of member, [T] =

      0   0  -1   0
      0   0   0  -1
      1   0   0   0
      0   1   0   0

Transformation matrix Transpose, [T] =

      0   0   1   0
      0   0   0   1
     -1   0   0   0
      0  -1   0   0

Global Matrix, [K global] =

      0.0013  -0.0013   0   0
     -0.0013   0.0013   0   0
           0        0   0   0
           0        0   0   0

Member Number = 4
Local Stiffness matrix of member, [K] =

      0   0        0        0
      0   0        0        0
      0   0   0.0011  -0.0011
      0   0  -0.0011   0.0011

Transformation matrix of member, [T] =

     0.7071        0   0.7071        0
          0   0.7071        0   0.7071
    -0.7071        0   0.7071        0
          0  -0.7071        0   0.7071
```

Transformation matrix Transpose, [T] =

```
    0.7071         0   -0.7071         0
         0    0.7071         0   -0.7071
    0.7071         0    0.7071         0
         0    0.7071         0    0.7071
```

Global Matrix, [K global] =

```
    1.0e-03 *
      0.5313   -0.5313   -0.5313    0.5313
     -0.5313    0.5313    0.5313   -0.5313
     -0.5313    0.5313    0.5313   -0.5313
      0.5313   -0.5313   -0.5313    0.5313
```

Member Number = 5
Local Stiffness matrix of member, [K] =

```
    0    0         0         0
    0    0         0         0
    0    0    0.0011   -0.0011
    0    0   -0.0011    0.0011
```

Transformation matrix of member, [T] =

```
    0.7071         0   -0.7071         0
         0    0.7071         0   -0.7071
    0.7071         0    0.7071         0
         0    0.7071         0    0.7071
```

Transformation matrix Transpose, [T] =

```
    0.7071         0    0.7071         0
         0    0.7071         0    0.7071
   -0.7071         0    0.7071         0
         0   -0.7071         0    0.7071
```

Global Matrix, [K global] =

```
    1.0e-03 *
      0.5313   -0.5313    0.5313   -0.5313
     -0.5313    0.5313   -0.5313    0.5313
      0.5313   -0.5313    0.5313   -0.5313
     -0.5313    0.5313   -0.5313    0.5313
```

Stiffness Matrix of complete structure, [Ktotal] =

```
    0.0015  -0.0005  -0.0010         0         0         0  -0.0005   0.0005
   -0.0005   0.0018         0         0         0  -0.0013   0.0005  -0.0005
   -0.0010        0   0.0015   0.0005  -0.0005  -0.0005         0         0
         0        0   0.0005   0.0018  -0.0005  -0.0005         0  -0.0013
         0        0  -0.0005  -0.0005   0.0005   0.0005         0         0
         0  -0.0013  -0.0005  -0.0005   0.0005   0.0018         0         0
   -0.0005   0.0005         0         0         0         0   0.0005  -0.0005
    0.0005  -0.0005         0  -0.0013         0         0  -0.0005   0.0018
```

Planar Truss Structures

```
Unrestrained Stiffness sub-matrix, [Kuu] =

    0.0015   -0.0005   -0.0010        0
   -0.0005    0.0018         0        0
   -0.0010         0    0.0015   0.0005
         0         0    0.0005   0.0018

Inverse of Unrestrained Stiffness sub-matrix, [KuuInverse] =

    1.0e+03 *
    1.5517    0.4629    1.1303   -0.3371
    0.4629    0.6994    0.3371   -0.1006
    1.1303    0.3371    1.5517   -0.4629
   -0.3371   -0.1006   -0.4629    0.6994

Joint Load vector, [Jl] =

    40  0  0  -20  0  0  0  0

Unrestrained displacements, [DelU] =

    1.0e+04 *
     6.8812    2.0525    5.4468   -2.7475

Member Number = 1
Global displacement matrix [DeltaBar] =

    1.0e+04 *
         0    2.0525    0    6.8812

Global End moment matrix [MBar] =

   -25.6568   25.6568    0    0

Member Number = 2
Global displacement matrix [DeltaBar] =

    1.0e+04 *
     2.0525   -2.7475    6.8812   5.4468

Global End moment matrix [MBar] =

    0    0    14.3432   -14.3432

Member Number = 3
Global displacement matrix [DeltaBar] =

    1.0e+04 *
         0   -2.7475    0    5.4468

Global End moment matrix [MBar] =

    34.3432   -34.3432    0    0
```

```
Member Number = 4
Global displacement matrix [DeltaBar] =

     1.0e+04 *
        2.0525      0    6.8812      0

Global End moment matrix [MBar] =

     -25.6568   25.6568   25.6568   -25.6568

Member Number = 5
Global displacement matrix [DeltaBar] =

     1.0e+04 *
            0   -2.7475     0    5.4468

Global End moment matrix [MBar] =

     -14.3432   14.3432   -14.3432   14.3432

Joint forces =

  40.0000  0.0000  -0.0000  -20.0000  -14.3432  -40.0000  -25.6568  60.0000
```

EXAMPLE 3.2:

Analyze the planar truss system shown in Figure 3.6 using the stiffness method. E is constant for all the members.

SOLUTION:

1. *Marking restrained and unrestrained degrees-of-freedom:*

    Unrestrained degrees-of-freedom = 4 (δ_1, δ_2, δ_3, δ_4, δ_5, δ_6, δ_7, δ_8)
    Restrained degrees-of-freedom = 4 (δ_9, δ_{10}, δ_{11}, δ_{12})

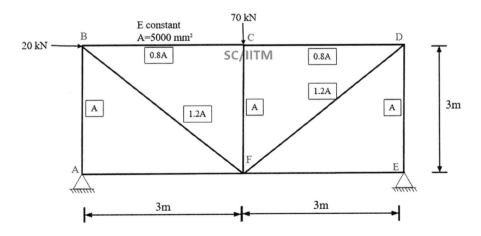

FIGURE 3.6 Truss example.

Planar Truss Structures

The unrestrained and restrained degrees-of-freedom are marked in the truss, as shown in Figure 3.7.

Member	Ends j	Ends k	Length (m)	θ (degrees)	C_x	C_y	Global Labels
AB	A	B	3	90	0	1	(10,2,9,1)
BC	B	C	3	0	1	0	(2,4,1,3)
CD	C	D	3	0	1	0	(4,6,3,5)
DE	E	D	3	90	0	1	(12,6,11,5)
AF	A	F	3	0	1	0	(10,8,9,7)
FE	F	E	3	0	1	0	(8,12,7,11)
BF	B	F	4.242	−45	0.707	−0.707	(2,8,1,7)
CF	C	F	3	+90	0	1	(8,4,7,3)
DF	D	F	4.242	−135	−0.707	0.707	(6,8,5,7)

2. *Calculation of the local stiffness matrix:*

   The local stiffness matrix for the truss element is given by,

$$[K_T]_i = \begin{bmatrix} 0 & 0 & 0 & 0 \\ 0 & 0 & 0 & 0 \\ 0 & 0 & \dfrac{AE}{l} & -\dfrac{AE}{l} \\ 0 & 0 & -\dfrac{AE}{l} & \dfrac{AE}{l} \end{bmatrix}$$

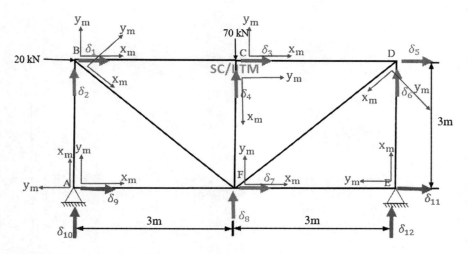

FIGURE 3.7 Marking restrained and unrestrained degrees-of-freedom.

$$K_{AB} = E \times 10^{-3} \begin{bmatrix} 0 & 0 & 0 & 0 \\ 0 & 0 & 0 & 0 \\ 0 & 0 & 1.7 & -1.7 \\ 0 & 0 & -1.7 & 1.7 \end{bmatrix}, \quad K_{BC} = E \times 10^{-3} \begin{bmatrix} 0 & 0 & 0 & 0 \\ 0 & 0 & 0 & 0 \\ 0 & 0 & 1.3 & -1.3 \\ 0 & 0 & -1.3 & 1.3 \end{bmatrix}$$

$$K_{CD} = E \times 10^{-3} \begin{bmatrix} 0 & 0 & 0 & 0 \\ 0 & 0 & 0 & 0 \\ 0 & 0 & 1.3 & -1.3 \\ 0 & 0 & -1.3 & 1.3 \end{bmatrix}, \quad K_{DE} = E \times 10^{-3} \begin{bmatrix} 0 & 0 & 0 & 0 \\ 0 & 0 & 0 & 0 \\ 0 & 0 & 1.7 & -1.7 \\ 0 & 0 & -1.7 & 1.7 \end{bmatrix}$$

$$K_{AF} = E \times 10^{-3} \begin{bmatrix} 0 & 0 & 0 & 0 \\ 0 & 0 & 0 & 0 \\ 0 & 0 & 1.3 & -1.3 \\ 0 & 0 & -1.3 & 1.3 \end{bmatrix}, \quad K_{FE} = E \times 10^{-3} \begin{bmatrix} 0 & 0 & 0 & 0 \\ 0 & 0 & 0 & 0 \\ 0 & 0 & 1.3 & -1.3 \\ 0 & 0 & -1.3 & 1.3 \end{bmatrix}$$

$$K_{BF} = E \times 10^{-3} \begin{bmatrix} 0 & 0 & 0 & 0 \\ 0 & 0 & 0 & 0 \\ 0 & 0 & 1.4 & -1.4 \\ 0 & 0 & -1.4 & 1.4 \end{bmatrix}, \quad K_{CF} = E \times 10^{-3} \begin{bmatrix} 0 & 0 & 0 & 0 \\ 0 & 0 & 0 & 0 \\ 0 & 0 & 1.7 & -1.7 \\ 0 & 0 & -1.7 & 1.7 \end{bmatrix}$$

$$K_{DF} = E \times 10^{-3} \begin{bmatrix} 0 & 0 & 0 & 0 \\ 0 & 0 & 0 & 0 \\ 0 & 0 & 1.4 & -1.4 \\ 0 & 0 & -1.4 & 1.4 \end{bmatrix}$$

3. Calculation of transformation matrix:

$$[T]_{AB} = \begin{bmatrix} 0 & 0 & -1 & 0 \\ 0 & 0 & 0 & -1 \\ 1 & 0 & 0 & 0 \\ 0 & 1 & 0 & 0 \end{bmatrix}, \quad [T]_{BC} = \begin{bmatrix} 1 & 0 & 0 & 0 \\ 0 & 1 & 0 & 0 \\ 0 & 0 & 1 & 0 \\ 0 & 0 & 0 & 1 \end{bmatrix}, \quad [T]_{CD} = \begin{bmatrix} 1 & 0 & 0 & 0 \\ 0 & 1 & 0 & 0 \\ 0 & 0 & 1 & 0 \\ 0 & 0 & 0 & 1 \end{bmatrix}$$

$$[T]_{DE} = \begin{bmatrix} 0 & 0 & -1 & 0 \\ 0 & 0 & 0 & -1 \\ 1 & 0 & 0 & 0 \\ 0 & 1 & 0 & 0 \end{bmatrix}, \quad [T]_{AF} = \begin{bmatrix} 1 & 0 & 0 & 0 \\ 0 & 1 & 0 & 0 \\ 0 & 0 & 1 & 0 \\ 0 & 0 & 0 & 1 \end{bmatrix}, \quad [T]_{FE} = \begin{bmatrix} 1 & 0 & 0 & 0 \\ 0 & 1 & 0 & 0 \\ 0 & 0 & 1 & 0 \\ 0 & 0 & 0 & 1 \end{bmatrix}$$

$$[T]_{BF} = \begin{bmatrix} 0.7071 & 0 & 0.7071 & 0 \\ 0 & 0.7071 & 0 & 0.7071 \\ -0.7071 & 0 & 0.7071 & 0 \\ 0 & -0.7071 & 0 & 0.7071 \end{bmatrix}, \quad [T]_{CF} = \begin{bmatrix} 0 & 0 & -1 & 0 \\ 0 & 0 & 0 & -1 \\ 1 & 0 & 0 & 0 \\ 0 & 1 & 0 & 0 \end{bmatrix}$$

Planar Truss Structures

$$[T]_{DF} = \begin{bmatrix} -0.7071 & 0 & 0.7071 & 0 \\ 0 & -0.7071 & 0 & 0.7071 \\ -0.7071 & 0 & -0.7071 & 0 \\ 0 & -0.7071 & 0 & -0.7071 \end{bmatrix}$$

4. *Calculation of the global stiffness matrix:*
   The global stiffness matrix of all the members are calculated by using the following relationship:

$$\left[\bar{K}_T\right]_i = \left[T_T\right]_i^T \left[K_T\right]_i \left[T_T\right]_i$$

$$\bar{K}_{AB} = E \begin{bmatrix} 0.0017 & -0.0017 & 0 & 0 \\ -0.0017 & 0.0017 & 0 & 0 \\ 0 & 0 & 0 & 0 \\ 0 & 0 & 0 & 0 \end{bmatrix} \begin{matrix} ⑩ \\ ② \\ ⑨ \\ ① \end{matrix}$$

(columns: ⑩ ② ⑨ ①)

$$\bar{K}_{BC} = E \begin{bmatrix} 0 & 0 & 0 & 0 \\ 0 & 0 & 0 & 0 \\ 0 & 0 & 0.0013 & -0.0013 \\ 0 & 0 & -0.0013 & 0.0013 \end{bmatrix} \begin{matrix} ② \\ ④ \\ ① \\ ③ \end{matrix}$$

(columns: ② ④ ① ③)

$$\bar{K}_{CD} = E \begin{bmatrix} 0 & 0 & 0 & 0 \\ 0 & 0 & 0 & 0 \\ 0 & 0 & 0.0013 & -0.0013 \\ 0 & 0 & -0.0013 & 0.0013 \end{bmatrix} \begin{matrix} ④ \\ ⑥ \\ ③ \\ ⑤ \end{matrix}$$

(columns: ④ ⑥ ③ ⑤)

$$\bar{K}_{DE} = E \begin{bmatrix} 0.0017 & -0.0017 & 0 & 0 \\ -0.0017 & 0.0017 & 0 & 0 \\ 0 & 0 & 0 & 0 \\ 0 & 0 & 0 & 0 \end{bmatrix} \begin{matrix} ⑫ \\ ⑥ \\ ⑪ \\ ⑤ \end{matrix}$$

(columns: ⑫ ⑥ ⑪ ⑤)

$$\bar{K}_{AF} = E \begin{bmatrix} 0 & 0 & 0 & 0 \\ 0 & 0 & 0 & 0 \\ 0 & 0 & 0.0013 & -0.0013 \\ 0 & 0 & -0.0013 & 0.0013 \end{bmatrix} \begin{matrix} ⑩ \\ ⑧ \\ ⑨ \\ ⑦ \end{matrix}$$

(columns: ⑩ ⑧ ⑨ ⑦)

$$\bar{K}_{FE} = E \begin{bmatrix} 0 & 0 & 0 & 0 \\ 0 & 0 & 0 & 0 \\ 0 & 0 & 0.0013 & -0.0013 \\ 0 & 0 & -0.0013 & 0.0013 \end{bmatrix} \begin{matrix} ⑧ \\ ⑫ \\ ⑦ \\ ⑪ \end{matrix}$$

(columns: ⑧ ⑫ ⑦ ⑪)

$$\bar{K}_{BF} = E \times 10^{-3} \begin{bmatrix} 0.7072 & -0.7072 & -0.7072 & 0.7072 \\ -0.7072 & 0.7072 & 0.7072 & -0.7072 \\ -0.7072 & 0.7072 & 0.7072 & -0.7072 \\ 0.7072 & -0.7072 & -0.7072 & 0.7072 \end{bmatrix} \begin{matrix} ② \\ ⑧ \\ ① \\ ⑦ \end{matrix}$$

(columns: ② ⑧ ① ⑦)

$$\bar{K}_{CF} = E \begin{bmatrix} 0.0017 & -0.0017 & 0 & 0 \\ -0.0017 & 0.0017 & 0 & 0 \\ 0 & 0 & 0 & 0 \\ 0 & 0 & 0 & 0 \end{bmatrix} \begin{matrix} ⑧ \\ ④ \\ ⑦ \\ ③ \end{matrix}$$

(columns: ⑧ ④ ⑦ ③)

$$\bar{K}_{DF} = E \times 10^{-3} \begin{bmatrix} 0.7072 & -0.7072 & 0.7072 & -0.7072 \\ -0.7072 & 0.7072 & -0.7072 & 0.7072 \\ 0.7072 & -0.7072 & 0.7072 & -0.7072 \\ -0.7072 & 0.7072 & -0.7072 & 0.7072 \end{bmatrix} \begin{matrix} ⑥ \\ ⑧ \\ ⑤ \\ ⑦ \end{matrix}$$

(columns: ⑥ ⑧ ⑤ ⑦)

The total stiffness matrix is obtained by assembling the global stiffness matrices of all members with respect to the global labels.
From the total stiffness matrix, the stiffness matrix for the unrestrained degrees-of-freedom can be partitioned.

$$[K_{uu}] = E \times 10^{-3} \begin{bmatrix} 2 & -0.7 & -1.3 & 0 & 0 & 0 & -0.7 & 0.7 \\ -0.7 & 2.4 & 0 & 0 & 0 & 0 & 0.7 & -0.7 \\ -1.3 & 0 & 2.7 & 0 & -1.3 & 0 & 0 & 0 \\ 0 & 0 & 0 & 1.7 & 0 & 0 & 0 & -1.7 \\ 0 & 0 & -1.3 & 0 & 1.2 & 0.7 & -0.7 & -0.7 \\ 0 & 0 & 0 & 0 & 0.7 & 2.4 & -0.7 & -0.7 \\ -0.7 & 0.7 & 0 & 0 & -0.7 & -0.7 & 4.1 & 0 \\ 0.7 & -0.7 & 0 & -1.7 & -0.7 & -0.7 & 0 & 3.1 \end{bmatrix} \begin{matrix} ① \\ ② \\ ③ \\ ④ \\ ⑤ \\ ⑥ \\ ⑦ \\ ⑧ \end{matrix}$$

(columns: ① ② ③ ④ ⑤ ⑥ ⑦ ⑧)

Planar Truss Structures

$$[K_{uu}]^{-1} = \frac{1}{E} \times 10^{-3} \begin{bmatrix} 1.757 & 0.3 & 1.382 & -0.375 & 1.007 & -0.3 & 0.375 & -0.375 \\ 0.3 & 0.6 & 0.3 & 0.3 & 1.3 & 0 & 0 & 0.3 \\ 1.382 & 0.3 & 1.7570 & 0 & 1.382 & -0.3 & 0.375 & 0 \\ -0.375 & 0.3 & 0 & 1.982 & 0.375 & 0.3 & 0 & 1.382 \\ 1.007 & 0.3 & 1.382 & 0.375 & 1.757 & -0.3 & 0.375 & 0.375 \\ -0.3 & 0 & -0.3 & 0.3 & -0.3 & 0.6 & 0 & 0.3 \\ 0.375 & 0 & 0.375 & 0 & 0.375 & 0 & 0.375 & 0 \\ -0.375 & 0.3 & 0 & 1.382 & 0.375 & 0.3 & 0 & 1.382 \end{bmatrix}$$

5. *Estimation of joint load vectors:*

The joint load vector can be written as follows:

$$[\bar{J}_t] = \begin{Bmatrix} 20 \\ 0 \\ 0 \\ -70 \\ 0 \\ 0 \\ 0 \\ 0 \\ 0 \\ 0 \\ 0 \\ 0 \end{Bmatrix} \begin{matrix} ① \\ ② \\ ③ \\ ④ \\ ⑤ \\ ⑥ \\ ⑦ \\ ⑧ \\ ⑨ \\ ⑩ \\ ⑪ \\ ⑫ \end{matrix}$$

The joint load vector in unrestrained degrees-of-freedom can be partitioned from the previous matrix as follows:

$$[\bar{J}_{Lu}] = \begin{Bmatrix} 20 \\ 0 \\ 0 \\ -70 \\ 0 \\ 0 \\ 0 \\ 0 \end{Bmatrix}$$

$$\{\bar{\delta}\} = [\bar{K}_{uu}]^{-1} \{\bar{J}_{Lu}\}$$

$$\{\bar{\delta}\}_U = \frac{1 \times 10^{-5}}{E} \begin{Bmatrix} 0.6139 \\ -0.15 \\ 0.2764 \\ -1.4624 \\ -0.0611 \\ -0.27 \\ 0.075 \\ -1.0424 \end{Bmatrix}$$

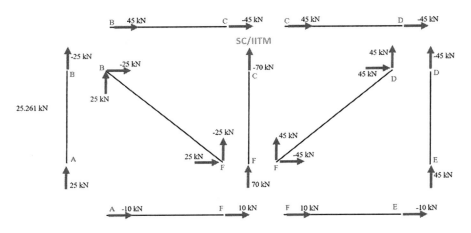

FIGURE 3.8 Member end forces.

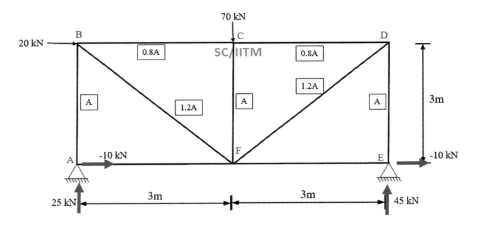

FIGURE 3.9 Final end forces.

6. *Calculation of member forces:*

$$\bar{M}_{AB} = \begin{Bmatrix} \bar{V}_{10} \\ \bar{V}_2 \\ \bar{H}_9 \\ \bar{H}_1 \end{Bmatrix} = \begin{Bmatrix} 25 \\ -25 \\ 0 \\ 0 \end{Bmatrix}, \quad \bar{M}_{BC} = \begin{Bmatrix} \bar{V}_2 \\ \bar{V}_4 \\ \bar{H}_1 \\ \bar{H}_3 \end{Bmatrix} = \begin{Bmatrix} 0 \\ 0 \\ 45 \\ -45 \end{Bmatrix}, \quad \bar{M}_{CD} = \begin{Bmatrix} \bar{V}_4 \\ \bar{V}_6 \\ \bar{H}_3 \\ \bar{H}_5 \end{Bmatrix} = \begin{Bmatrix} 0 \\ 0 \\ 45 \\ -45 \end{Bmatrix}$$

$$\bar{M}_{DE} = \begin{Bmatrix} \bar{V}_{12} \\ \bar{V}_6 \\ \bar{H}_{11} \\ \bar{H}_5 \end{Bmatrix} = \begin{Bmatrix} 45 \\ -45 \\ 0 \\ 0 \end{Bmatrix}, \quad \bar{M}_{AF} = \begin{Bmatrix} \bar{V}_{10} \\ \bar{V}_8 \\ \bar{H}_9 \\ \bar{H}_7 \end{Bmatrix} = \begin{Bmatrix} 0 \\ 0 \\ -10 \\ 10 \end{Bmatrix}, \quad \bar{M}_{FE} = \begin{Bmatrix} \bar{V}_8 \\ \bar{V}_{12} \\ \bar{H}_7 \\ \bar{H}_{11} \end{Bmatrix} = \begin{Bmatrix} 0 \\ 0 \\ 10 \\ -10 \end{Bmatrix}$$

Planar Truss Structures

$$\bar{M}_{BF} = \begin{Bmatrix} \bar{V}_2 \\ \bar{V}_8 \\ \bar{H}_1 \\ \bar{H}_7 \end{Bmatrix} = \begin{Bmatrix} 25 \\ -25 \\ -25 \\ 25 \end{Bmatrix}, \quad \bar{M}_{CF} = \begin{Bmatrix} \bar{V}_8 \\ \bar{V}_4 \\ \bar{H}_7 \\ \bar{H}_3 \end{Bmatrix} = \begin{Bmatrix} 70 \\ -70 \\ 0 \\ 0 \end{Bmatrix}, \quad \bar{M}_{DF} = \begin{Bmatrix} \bar{V}_6 \\ \bar{V}_8 \\ \bar{H}_5 \\ \bar{H}_7 \end{Bmatrix} = \begin{Bmatrix} 45 \\ -45 \\ 45 \\ -45 \end{Bmatrix}$$

The member end forces and final end forces are given in Figures 3.8 and 3.9 respectively.

MATLAB program:

```
%% stiffness matrix method for planar truss
% Input
clc;
clear;
n = 9; % number of members
L = [3 3 3 3 3 3 4.242 3 4.242]; % length in m
A = [5E-3 4E-3 4E-3 5E-3 4E-3 4E-3 6E-3 5E-3 6E-3]; % Area in m2
theta= [90 0 0 90 0 0 -45 90 -135]; % angle in degrees
uu = 8; % Number of unrestrained degrees-of-freedom
ur = 4; % Number of restrained degrees-of-freedom
uul = [1 2 3 4 5 6 7 8]; % global labels of unrestrained dof
url = [9 10 11 12]; % global labels of restrained dof
l1 = [10 2 9 1]; % Global labels for member 1
l2 = [2 4 1 3]; % Global labels for member 2
l3 = [4 6 3 5]; % Global labels for member 3
l4 = [12 6 11 5]; % Global labels for member 4
l5 = [10 8 9 7]; % Global labels for member 5
l6 = [8 12 7 11]; % Global labels for member 6
l7 = [2 8 1 7]; % Global labels for member 7
l8 = [8 4 7 3]; % Global labels for member 8
l9 = [6 8 5 7]; % Global labels for member 9
l= [l1; l2; l3; l4; l5; l6; l7; l8; l9];
dof = uu + ur; % Degrees-of-freedom
Ktotal = zeros (dof);
Tt1 = zeros (4); % Transformation matrix for member 1
Tt2 = zeros (4); % Transformation matrix for member 2
Tt3 = zeros (4); % Transformation matrix for member 3
Tt4 = zeros (4); % Transformation matrix for member 4
Tt5 = zeros (4); % Transformation matrix for member 5
Tt6 = zeros (4); % Transformation matrix for member 6
Tt7 = zeros (4); % Transformation matrix for member 7
Tt8 = zeros (4); % Transformation matrix for member 8
Tt9 = zeros (4); % Transformation matrix for member 9
fem1= [0; 0; 0; 0]; % Local Fixed end forces of member 1
fem2= [0; 0; 0; 0]; % Local Fixed end forces of member 2
fem3= [0; 0; 0; 0]; % Local Fixed end forces of member 3
fem4= [0; 0; 0; 0]; % Local Fixed end forces of member 4
fem5= [0; 0; 0; 0]; % Local Fixed end forces of member 5
fem6= [0; 0; 0; 0]; % Local Fixed end forces of member 6
fem7= [0; 0; 0; 0]; % Local Fixed end forces of member 7
```

```
fem8= [0; 0; 0; 0]; % Local Fixed end forces of member 8
fem9= [0; 0; 0; 0]; % Local Fixed end forces of member 9

%% rotation coefficients for each member
rc = A./L;
cx = cosd(theta);
cy = sind(theta);

%% stiffness matrix 4 by 4
for i = 1:n
      Knew = zeros (dof);
      k1 = [0; 0; 0; 0];
      k2 = [0; 0; 0; 0];
      k3 = [0; 0; rc(i); -rc(i)];
      k4 = -k3;
      K = [k1 k2 k3 k4];
      fprintf ('Member Number =');
      disp (i);
      fprintf ('Local Stiffness matrix of member, [K] = \n');
      disp (K);
      T1 = [cx(i); 0; cy(i); 0];
      T2 = [0; cx(i); 0; cy(i)];
      T3 = [-cy(i); 0; cx(i); 0];
      T4 = [0; -cy(i); 0; cx(i)];
      T = [T1 T2 T3 T4];
      fprintf ('Transformation matrix of member, [T] = \n');
      disp (T);
      Ttr = T';
      fprintf ('Transformation matrix Transpose, [T] = \n');
      disp (Ttr);
      Kg = Ttr*K*T;
      fprintf ('Global Matrix, [K global] = \n');
      disp (Kg);
      for p = 1:4
         for q = 1:4
            Knew((l(i,p)),(l(i,q))) =Kg(p,q);
         end
      end
      Ktotal = Ktotal + Knew;
      if i == 1
         Tt1= T;
         Kg1=Kg;
         fembar1= Tt1'*fem1;
      elseif i == 2
         Tt2 = T;
         Kg2 = Kg;
         fembar2= Tt2'*fem2;
      elseif i == 3
         Tt3 = T;
         Kg3 = Kg;
         fembar3= Tt3'*fem3;
```

Planar Truss Structures

```
            elseif i == 4
                Tt4 = T;
                Kg4 = Kg;
                fembar4= Tt4'*fem4;
            elseif i == 5
                Tt5 = T;
                Kg5 = Kg;
                fembar5= Tt5'*fem5;
            elseif i == 6
                Tt6 = T;
                Kg6 = Kg;
                fembar6= Tt6'*fem6;
            elseif i == 7
                Tt7 = T;
                Kg7 = Kg;
                fembar7= Tt7'*fem7;
            elseif i == 8
                Tt8 = T;
                Kg8 = Kg;
                fembar8= Tt8'*fem8;
            else
                Tt9 = T;
                Kg9=Kg;
                fembar9= Tt9'*fem9;
            end
end
fprintf ('Stiffness Matrix of complete structure, [Ktotal] = \n');
disp (Ktotal);
Kunr = zeros(uu);
for x=1:uu
    for y=1:uu
        Kunr(x,y)= Ktotal(x,y);
    end
end
fprintf ('Unrestrained Stiffness sub-matrix, [Kuu] = \n');
disp (Kunr);
KuuInv= inv(Kunr);
fprintf ('Inverse of Unrestrained Stiffness sub-matrix,
[KuuInverse] = \n');
disp (KuuInv);

%% Creation of joint load vector
jl= [20; 0; 0; -70; 0; 0; 0; 0; 0; 0; 0; 0]; % values given in kN
jlu = [20; 0; 0; -70; 0; 0; 0; 0]; % load vector in
unrestrained dof
delu = KuuInv*jlu;
fprintf ('Joint Load vector, [Jl] = \n');
disp (jl');
fprintf ('Unrestrained displacements, [DelU] = \n');
disp (delu');
delr = zeros(ur,1);
del = zeros (dof,1);
```

```
del = [delu; delr];
deli= zeros (4,1);
for i = 1:n
   for p = 1:4
       deli(p,1) = del((l(i,p)),1) ;
   end
   if i == 1
         delbar1 = deli;
         mbar1= (Kg1 * delbar1)+fembar1;
         fprintf ('Member Number =');
         disp (i);
         fprintf ('Global displacement matrix [DeltaBar] = \n');
         disp (delbar1');
         fprintf ('Global End moment matrix [MBar] = \n');
         disp (mbar1');
      elseif i == 2
         delbar2 = deli;
         mbar2= (Kg2 * delbar2)+fembar2;
         fprintf ('Member Number =');
         disp (i);
         fprintf ('Global displacement matrix [DeltaBar] = \n');
         disp (delbar2');
         fprintf ('Global End moment matrix [MBar] = \n');
         disp (mbar2');
      elseif i ==3
         delbar3 = deli;
         mbar3= (Kg3 * delbar3)+fembar3;
         fprintf ('Member Number =');
         disp (i);
         fprintf ('Global displacement matrix [DeltaBar] = \n');
         disp (delbar3');
         fprintf ('Global End moment matrix [MBar] = \n');
         disp (mbar3');
      elseif i == 4
         delbar4 = deli;
         mbar4= (Kg4 * delbar4)+fembar4;
         fprintf ('Member Number =');
         disp (i);
         fprintf ('Global displacement matrix [DeltaBar] = \n');
         disp (delbar4');
         fprintf ('Global End moment matrix [MBar] = \n');
         disp (mbar4');
      elseif i ==5
         delbar5 = deli;
         mbar5= (Kg5 * delbar5)+fembar5;
         fprintf ('Member Number =');
         disp (i);
         fprintf ('Global displacement matrix [DeltaBar] = \n');
         disp (delbar5');
         fprintf ('Global End moment matrix [MBar] = \n');
         disp (mbar5');
      elseif i ==6
```

Planar Truss Structures

```
            delbar6 = deli;
            mbar6= (Kg6 * delbar6)+fembar6;
            fprintf ('Member Number =');
            disp (i);
            fprintf ('Global displacement matrix [DeltaBar] = \n');
            disp (delbar6');
            fprintf ('Global End moment matrix [MBar] = \n');
            disp (mbar6');
        elseif i ==7
            delbar7 = deli;
            mbar7= (Kg7 * delbar7)+fembar7;
            fprintf ('Member Number =');
            disp (i);
            fprintf ('Global displacement matrix [DeltaBar] = \n');
            disp (delbar7');
            fprintf ('Global End moment matrix [MBar] = \n');
            disp (mbar7');
        elseif i ==8
            delbar8 = deli;
            mbar8= (Kg8 * delbar8)+fembar8;
            fprintf ('Member Number =');
            disp (i);
            fprintf ('Global displacement matrix [DeltaBar] = \n');
            disp (delbar8');
            fprintf ('Global End moment matrix [MBar] = \n');
            disp (mbar8');
        else
            delbar9 = deli;
            mbar9= (Kg9 * delbar9)+fembar9;
            fprintf ('Member Number =');
            disp (i);
            fprintf ('Global displacement matrix [DeltaBar] = \n');
            disp (delbar9');
            fprintf ('Global End moment matrix [MBar] = \n');
            disp (mbar9');
        end
end

%% check
mbar = [mbar1'; mbar2'; mbar3'; mbar4'; mbar5'; mbar6';
mbar7'; mbar8'; mbar9'];
jf = zeros(dof,1);
for a=1:n
    for b=1:4 % size of k matrix
        d = l(a,b);
        jfnew = zeros(dof,1);
        jfnew(d,1)=mbar(a,b);
        jf=jf+jfnew;
    end
end
fprintf ('Joint forces = \n');
disp (jf');
```

MATLAB output:

```
Member Number = 1
Local Stiffness matrix of member, [K] =

    0    0       0         0
    0    0       0         0
    0    0    0.0017   -0.0017
    0    0   -0.0017    0.0017

Transformation matrix of member, [T] =

    0    0   -1    0
    0    0    0   -1
    1    0    0    0
    0    1    0    0

Transformation matrix Transpose, [T] =

    0    0    1    0
    0    0    0    1
   -1    0    0    0
    0   -1    0    0

Global Matrix, [K global] =

    0.0017   -0.0017    0    0
   -0.0017    0.0017    0    0
         0         0    0    0
         0         0    0    0

Member Number = 2
Local Stiffness matrix of member, [K] =

    0    0       0         0
    0    0       0         0
    0    0    0.0013   -0.0013
    0    0   -0.0013    0.0013

Transformation matrix of member, [T] =

    1    0    0    0
    0    1    0    0
    0    0    1    0
    0    0    0    1

Transformation matrix Transpose, [T] =

    1    0    0    0
    0    1    0    0
    0    0    1    0
    0    0    0    1
```

Planar Truss Structures

Global Matrix, [K global] =

```
0  0      0         0
0  0      0         0
0  0   0.0013   -0.0013
0  0  -0.0013    0.0013
```

Member Number = 3
Local Stiffness matrix of member, [K] =

```
0  0      0         0
0  0      0         0
0  0   0.0013   -0.0013
0  0  -0.0013    0.0013
```

Transformation matrix of member, [T] =

```
1  0  0  0
0  1  0  0
0  0  1  0
0  0  0  1
```

Transformation matrix Transpose, [T] =

```
1  0  0  0
0  1  0  0
0  0  1  0
0  0  0  1
```

Global Matrix, [K global] =

```
0  0      0         0
0  0      0         0
0  0   0.0013   -0.0013
0  0  -0.0013    0.0013
```

Member Number = 4
Local Stiffness matrix of member, [K] =

```
0  0      0         0
0  0      0         0
0  0   0.0017   -0.0017
0  0  -0.0017    0.0017
```

Transformation matrix of member, [T] =

```
0  0  -1   0
0  0   0  -1
1  0   0   0
0  1   0   0
```

Transformation matrix Transpose, [T] =

```
 0   0   1   0
 0   0   0   1
-1   0   0   0
 0  -1   0   0
```

Global Matrix, [K global] =

```
 0.0017  -0.0017   0   0
-0.0017   0.0017   0   0
      0        0   0   0
      0        0   0   0
```

Member Number = 5
Local Stiffness matrix of member, [K] =

```
0   0    0        0
0   0    0        0
0   0    0.0013  -0.0013
0   0   -0.0013   0.0013
```

Transformation matrix of member, [T] =

```
1   0   0   0
0   1   0   0
0   0   1   0
0   0   0   1
```

Transformation matrix Transpose, [T] =

```
1   0   0   0
0   1   0   0
0   0   1   0
0   0   0   1
```

Global Matrix, [K global] =

```
0   0    0        0
0   0    0        0
0   0    0.0013  -0.0013
0   0   -0.0013   0.0013
```

Member Number =   6
Local Stiffness matrix of member, [K] =

```
0   0    0        0
0   0    0        0
0   0    0.0013  -0.0013
0   0   -0.0013   0.0013
```

Planar Truss Structures

Transformation matrix of member, [T] =

```
1  0  0  0
0  1  0  0
0  0  1  0
0  0  0  1
```

Transformation matrix Transpose, [T] =

```
1  0  0  0
0  1  0  0
0  0  1  0
0  0  0  1
```

Global Matrix, [K global] =

```
0  0   0        0
0  0   0        0
0  0   0.0013  -0.0013
0  0  -0.0013   0.0013
```

Member Number = 7
Local Stiffness matrix of member, [K] =

```
0  0   0        0
0  0   0        0
0  0   0.0014  -0.0014
0  0  -0.0014   0.0014
```

Transformation matrix of member, [T] =

```
 0.7071    0       0.7071    0
 0         0.7071  0         0.7071
-0.7071    0       0.7071    0
 0        -0.7071  0         0.7071
```

Transformation matrix Transpose, [T] =

```
 0.7071    0      -0.7071    0
 0         0.7071  0        -0.7071
 0.7071    0       0.7071    0
 0         0.7071  0         0.7071
```

Global Matrix, [K global] =

1.0e-03 *

```
 0.7072  -0.7072  -0.7072   0.7072
-0.7072   0.7072   0.7072  -0.7072
-0.7072   0.7072   0.7072  -0.7072
 0.7072  -0.7072  -0.7072   0.7072
```

```
Member Number = 8
Local Stiffness matrix of member, [K] =

    0    0       0         0
    0    0       0         0
    0    0    0.0017   -0.0017
    0    0   -0.0017    0.0017

Transformation matrix of member, [T] =

    0    0   -1    0
    0    0    0   -1
    1    0    0    0
    0    1    0    0

Transformation matrix Transpose, [T] =

    0    0    1    0
    0    0    0    1
   -1    0    0    0
    0   -1    0    0

Global Matrix, [K global] =

    0.0017   -0.0017   0   0
   -0.0017    0.0017   0   0
         0         0   0   0
         0         0   0   0

Member Number = 9
Local Stiffness matrix of member, [K] =

    0    0       0         0
    0    0       0         0
    0    0    0.0014   -0.0014
    0    0   -0.0014    0.0014

Transformation matrix of member, [T] =

   -0.7071        0     0.7071        0
         0   -0.7071         0    0.7071
   -0.7071        0    -0.7071        0
         0   -0.7071         0   -0.7071

Transformation matrix Transpose, [T] =

   -0.7071        0    -0.7071        0
         0   -0.7071         0   -0.7071
    0.7071        0    -0.7071        0
         0    0.7071         0   -0.7071
```

Planar Truss Structures

```
Global Matrix, [K global] =

    1.0e-03 *
     0.7072    -0.7072     0.7072    -0.7072
    -0.7072     0.7072    -0.7072     0.7072
     0.7072    -0.7072     0.7072    -0.7072
    -0.7072     0.7072    -0.7072     0.7072

Stiffness Matrix of complete structure, [Ktotal] =
Columns 1 through 9

  0.0020 -0.0007 -0.0013       0       0       0 -0.0007  0.0007       0
 -0.0007  0.0024       0       0       0       0  0.0007 -0.0007       0
 -0.0013       0  0.0027       0 -0.0013       0       0       0       0
       0       0       0  0.0017       0       0       0 -0.0017       0
       0       0 -0.0013       0  0.0020  0.0007 -0.0007 -0.0007       0
       0       0       0       0  0.0007  0.0024 -0.0007 -0.0007       0
 -0.0007  0.0007       0       0 -0.0007 -0.0007  0.0041       0 -0.0013
  0.0007 -0.0007       0 -0.0017 -0.0007 -0.0007       0  0.0031       0
       0       0       0       0       0       0 -0.0013       0  0.0013
       0 -0.0017       0       0       0       0       0       0       0
       0       0       0       0       0       0 -0.0013       0       0
       0       0       0       0       0 -0.0017       0       0       0

Columns 10 through 12

            0           0           0
      -0.0017           0           0
            0           0           0
            0           0           0
            0           0           0
            0           0     -0.0017
            0     -0.0013           0
            0           0           0
            0           0           0
       0.0017           0           0
            0      0.0013           0
            0           0      0.0017

Unrestrained Stiffness sub-matrix, [Kuu] =

   0.0020 -0.0007 -0.0013       0       0       0 -0.0007  0.0007
  -0.0007  0.0024       0       0       0       0  0.0007 -0.0007
  -0.0013       0  0.0027       0 -0.0013       0       0       0
        0       0       0  0.0017       0       0       0 -0.0017
        0       0 -0.0013       0  0.0020  0.0007 -0.0007 -0.0007
        0       0       0       0  0.0007  0.0024 -0.0007 -0.0007
  -0.0007  0.0007       0       0 -0.0007 -0.0007  0.0041       0
   0.0007 -0.0007       0 -0.0017 -0.0007 -0.0007       0  0.0031
```

156 Advanced Structural Analysis with MATLAB®

```
Inverse of Unrestrained Stiffness sub-matrix, [KuuInverse] =

 1.0e+03 *
  1.7570    0.3000    1.3820   -0.3750    1.0070   -0.3000    0.3750   -0.3750
  0.3000    0.6000    0.3000    0.3000    0.3000         0    0.0000    0.3000
  1.3820    0.3000    1.7570   -0.0000    1.3820   -0.3000    0.3750   -0.0000
 -0.3750    0.3000   -0.0000    1.9820    0.3750    0.3000   -0.0000    1.3820
  1.0070    0.3000    1.3820    0.3750    1.7570   -0.3000    0.3750    0.3750
 -0.3000         0   -0.3000    0.3000   -0.3000    0.6000   -0.0000    0.3000
  0.3750    0.0000    0.3750   -0.0000    0.3750   -0.0000    0.3750   -0.0000
 -0.3750    0.3000   -0.0000    1.3820    0.3750    0.3000   -0.0000    1.3820

Joint Load vector, [Jl] =

    20     0     0   -70     0     0     0     0     0     0     0     0

Unrestrained displacements, [DelU] =

 1.0e+05 *
  0.6139   -0.1500  0.2764  -1.4624  -0.0611  -0.2700  0.0750  -1.0424

Member Number = 1
Global displacement matrix [DeltaBar] =

    1.0e+04 *
         0    -1.5000     0    6.1390

Global End moment matrix [MBar] =

    25.0000   -25.0000    0    0

Member Number = 2
Global displacement matrix [DeltaBar] =

    1.0e+05 *
    -0.1500    -1.4624    0.6139    0.2764

Global End moment matrix [MBar] =

    0    0    45.0000   -45.0000

Member Number = 3
Global displacement matrix [DeltaBar] =

    1.0e+05 *
    -1.4624    -0.2700    0.2764    -0.0611

Global End moment matrix [MBar] =

    0    0    45.0000   -45.0000

Member Number = 4
Global displacement matrix [DeltaBar] =

    1.0e+04 *
         0    -2.7000     0    -0.6110
```

Planar Truss Structures

```
Global End moment matrix [MBar] =

    45.0000   -45.0000   0   0

Member Number = 5
Global displacement matrix [DeltaBar] =

    1.0e+05 *
         0    -1.0424   0   0.0750

Global End moment matrix [MBar] =

    0   0   -10.0000   10.0000

Member Number = 6
Global displacement matrix [DeltaBar] =

    1.0e+05 *
    -1.0424   0   0.0750   0

Global End moment matrix [MBar] =

    0   0   10.0000   -10.0000

Member Number = 7
Global displacement matrix [DeltaBar] =

    1.0e+05 *
    -0.1500    -1.0424    0.6139    0.0750

Global End moment matrix [MBar] =

    25.0000   -25.0000   -25.0000   25.0000

Member Number = 8
Global displacement matrix [DeltaBar] =

    1.0e+05 *
    -1.0424    -1.4624    0.0750    0.2764

Global End moment matrix [MBar] =

    70   -70   0   0

Member Number = 9
Global displacement matrix [DeltaBar] =

    1.0e+05 *
    -0.2700    -1.0424    -0.0611    0.0750

Global End moment matrix [MBar] =

    45.0000   -45.0000   45.0000   -45.0000

Joint forces =

  20.0000    0.0000    0.0000   -70.0000  0  0.0000  -0.0000  -0.0000  -10.0000
  25.0000  -10.0000  45.0000
```

4 Three-Dimensional Analysis of Space Frames

4.1 THREE-DIMENSIONAL ANALYSIS OF STRUCTURES

In the previous section, the detailed explanation on the stiffness method of analysis of planar orthogonal and non-orthogonal structures using beam elements and planar truss elements were given. The same algorithm and sign convention are now extended to solve the three-dimensional structures, in order to make the analysis very simple and easier. Let us extend the basics of the beam element discussed so far to the three-dimensional structures. We know that the equation for joint equilibrium of the planar structure is given by,

$$[K]_{\text{complete}}\{\Delta\}_{\text{complete}} = \{J_L\}_{\text{complete}} + \{R\}_{\text{complete}} \tag{4.1}$$

The previous equation is also expandable to solve three-dimensional structures. Similarly, the matrix equation describing equilibrium of the beam element is given by,

$$\{M\}_i = [K_i][T_i]\{\delta_i\} + \{\text{FEM}\}_i \tag{4.2}$$

This equation can also be extended to analyze three-dimensional structures consisting of beam elements arbitrarily oriented in space.

The first task in three-dimensional analysis is to develop the stiffness matrix of the complete structure, which can be simply done by the summation of member stiffness matrices of individual elements. It is very important to note that the complete stiffness matrix of the space system will be established in the reference axes system.

Sign convention:

Consider three axes, as shown in Figure 4.1, where the vector representing the translation or forces are marked. Then, the right-hand thumb rule is used to mark the direction of moment in every axis. If the thumb points toward the arrows, then the direction of the remaining four fingers will indicate the direction of moment. This indicates the direction of rotation or moment. All these are considered to be positively established using the right-hand thumb rule of orthogonal coordinate axes.

4.2 BEAM ELEMENT

Let us consider a beam element fixed at both ends with the local axes (x_m-y_m), as shown in Figure 4.2. The beam has six degrees-of-freedom.

Let us extend the same algorithm used in the analysis of planar orthogonal structure to the three-dimensional structure arbitrarily oriented in space. Consider a member with two joints 'j' and 'k', as shown in Figure 4.3. The local axes system is

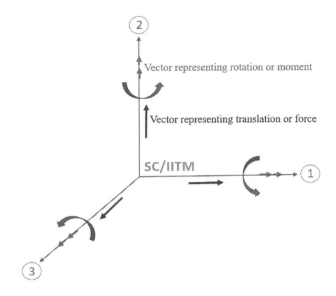

FIGURE 4.1 Vector representation for translation and rotation.

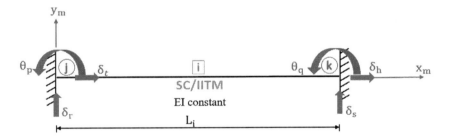

FIGURE 4.2 Fixed beam element.

represented by $(x_m\text{-}y_m\text{-}z_m)$, which are orthogonal to each other. The reference axes system is represented by (X-Y-Z). The degrees-of-freedom are also marked in the same manner, similar to that of the two-dimensional members. At every end, there are three translations along three directions and three rotations about three dimensions. Thus, there are twelve degrees-of-freedom now.

At the jth end, translations are (t,r,v) and rotations are (l,n,p). Similarly, at the kth end, the translations are (h,s,w) and rotations are (m,o,q). Thus, the stiffness matrix will be of size 12×12.

4.3 THE STIFFNESS MATRIX

Let us consider unit displacement or translation along 't' in x_m direction at the jth end, as shown in Figure 4.4. Similarly, unit translation is given in y_m direction at the jth end. In these two cases, the translation occurs in the (x-y) plane. The unit

Three-Dimensional Analysis of Space Frames

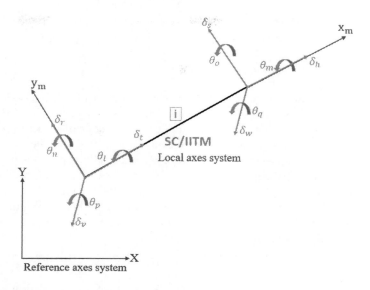

FIGURE 4.3 Degrees-of-freedom in local axes system.

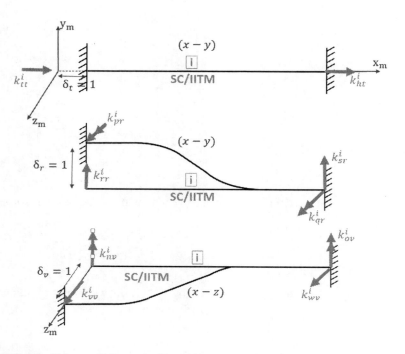

FIGURE 4.4 Unit translation at the jth end in x_m-y_m-z_m directions.

translation is also given in the z_m axis direction and it happens in the (x-z) plane. The corresponding stiffness coefficients due to the applied unit translation are also marked.

Let us apply unit rotations for the member, as shown in Figure 4.5, at the *j*th end in all three axes.

The same procedure can be extended for the *k*th end also. Let us develop the coefficients contributing the stiffness matrix. When unit displacement is given along 't', only t and h are influenced, and the remaining coefficients will be zero. Thus, the first column of the stiffness matrix will have only two stiffness coefficients: k_{tt}, k_{ht}. In the same way, the stiffness coefficients are entered in all the columns by finding the invoked or influenced degrees-of-freedom under the application of unit displacement or unit rotation. The stiffness matrix is shown in Figure 4.6.

Now, the values of these stiffness coefficients have to be found from the known principal quantities of the beam element, in order to get the complete stiffness matrix of a three-dimensional beam element. Thus, the complete stiffness matrix is given by,

$$[K] = \begin{bmatrix} \frac{EA_x}{l} & 0 & 0 & 0 & 0 & 0 & -\frac{EA_x}{l} & 0 & 0 & 0 & 0 & 0 \\ 0 & \frac{12EI_z}{l^3} & 0 & 0 & 0 & \frac{6EI_z}{l^2} & 0 & -\frac{12EI_z}{l^3} & 0 & 0 & 0 & \frac{6EI_z}{l^2} \\ 0 & 0 & \frac{12EI_y}{l^3} & 0 & -\frac{6EI_y}{l^2} & 0 & 0 & 0 & -\frac{12EI_y}{l^3} & 0 & -\frac{6EI_y}{l^2} & 0 \\ 0 & 0 & 0 & \frac{GI_x}{l} & 0 & 0 & 0 & 0 & 0 & -\frac{GI_x}{l} & 0 & 0 \\ 0 & 0 & -\frac{6EI_y}{l^2} & 0 & \frac{4EI_y}{l} & 0 & 0 & 0 & \frac{6EI_y}{l^2} & 0 & \frac{2EI_y}{l} & 0 \\ 0 & \frac{6EI_z}{l^2} & 0 & 0 & 0 & \frac{4EI_z}{l} & 0 & -\frac{6EI_z}{l^2} & 0 & 0 & 0 & \frac{2EI_z}{l} \\ -\frac{EA_x}{l} & 0 & 0 & 0 & 0 & 0 & \frac{EA_x}{l} & 0 & 0 & 0 & 0 & 0 \\ 0 & -\frac{12EI_z}{l^3} & 0 & 0 & 0 & -\frac{6EI_z}{l^2} & 0 & \frac{12EI_z}{l^3} & 0 & 0 & 0 & -\frac{6EI_z}{l^2} \\ 0 & 0 & -\frac{12EI_y}{l^3} & 0 & \frac{6EI_y}{l^2} & 0 & 0 & 0 & \frac{12EI_y}{l^3} & 0 & \frac{6EI_y}{l^2} & 0 \\ 0 & 0 & 0 & -\frac{GI_x}{l} & 0 & 0 & 0 & 0 & 0 & \frac{GI_x}{l} & 0 & 0 \\ 0 & 0 & -\frac{6EI_y}{l^2} & 0 & \frac{2EI_y}{l} & 0 & 0 & 0 & \frac{6EI_y}{l^2} & 0 & \frac{4EI_y}{l} & 0 \\ 0 & \frac{6EI_z}{l^2} & 0 & 0 & 0 & \frac{2EI_z}{l} & 0 & -\frac{6EI_z}{l^2} & 0 & 0 & 0 & \frac{4EI_z}{l} \end{bmatrix}$$

I_x is called torsional constant. For the beam element, with rectangular cross-section, I_x is given by,

$$I_x = \frac{ht^3}{12}\left[1 - 0.63\left(\frac{t}{h}\right) + 0.052\left(\frac{t}{h}\right)^3\right] \quad \text{for} \quad h \gg t \quad (4.3)$$

where, h is the depth of the rectangular section and t is the width of the section. For a rectangular cross-section with a very large value of h/t,

$$I_x = \frac{ht^3}{12} \quad (4.4)$$

Three-Dimensional Analysis of Space Frames

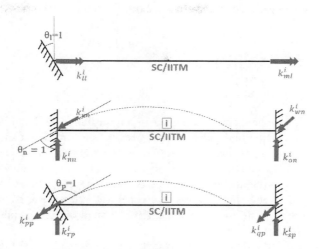

FIGURE 4.5 Unit rotation at the jth end in x_m-y_m-z_m directions.

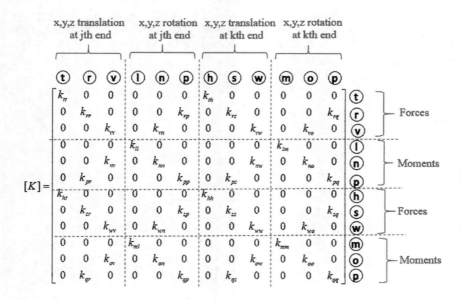

FIGURE 4.6 Stiffness coefficients.

In case of I-sections, as shown in Figure 4.7, the value of torsional constant is given by,

$$I_x = \frac{1}{3}\sum ht^3 \tag{4.5}$$

where, h is the longer dimension and t is the shorter dimension. Hence,

$$I_x = \frac{1}{3}\left\{\left(b_f t_f^3\right) + \left(h_w t_w^3\right) + \left(b_f t_f^3\right)\right\} \qquad (4.6)$$

4.4 TRANSFORMATION MATRIX

In a space frame, members can be oriented in any fashion. The local axes of the member may not coincide with the reference axes system. In such a situation, the stiffness matrix needs to be transformed with respect to the reference axes system. In addition, the load applied on the local axes also needs to be transformed with respect to the reference axes system. Most importantly, the member forces, end moments and reactions need to be computed with respect to the reference axes system, but they also need to be transformed to the local axes system of each member. This is required to design the member.

Let V_0 be a vector that is arbitrarily oriented along the axis Y_0. This has to be transformed to the reference axes system (Y_1-Y_2-Y_3). The vector V_0 has its components along the Y_1, Y_2 and Y_3 axes. In order to find the components, the inclination or position of the vector with respect to the three axes should be known. In Figure 4.8, γ_{01} is the angle between the axes Y_0 and Y_1. Similarly, γ_{02} is the angle between the axes Y_0 and Y_2 and γ_{03} is the angle between the axes Y_0 and Y_3. The corresponding components are V_1, V_2 and V_3. Thus, the following relationship will be valid.

$$V_1 = V_0 \cos \gamma_{01}$$
$$V_2 = V_0 \cos \gamma_{02} \qquad (4.7)$$
$$V_3 = V_0 \cos \gamma_{03}$$

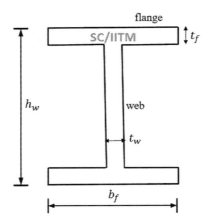

FIGURE 4.7 I section.

Three-Dimensional Analysis of Space Frames

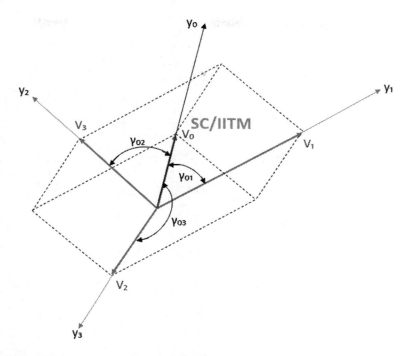

FIGURE 4.8 Component of vector along coordinate axes.

where, γ_{01}, γ_{02} and γ_{03} are defined as angles between the vector V_0 or the axis Y_0 to which the vector is aligned, whose coordinate axes are $(Y_1\text{-}Y_2\text{-}Y_3)$ respectively. In the previous equation, the terms $\cos\gamma_{01}$, $\cos\gamma_{02}$ and $\cos\gamma_{03}$ are called direction cosines. Now, the vector V_0 is resolved along the set of coordinate axes $(Y_1\text{-}Y_2\text{-}Y_3)$. Let us resolve or transform this to the standard reference axes system (X-Y-Z).

Let, $X_1\text{-}X_2\text{-}X_3$ be the reference axes systems. Let us define the angles of $X_1\text{-}X_2\text{-}X_3$ and vectors V_1, V_2, V_3. V_1, V_2, V_3 are already resolved along $(Y_1\text{-}Y_2\text{-}Y_3)$. Let, γ_{11} be the angle between the V_1 axis and X_1 axis. By this logic, the angle between the vector and the reference axes system can be defined. Thus, the V_1 vector makes an angle $(\gamma_{11}, \gamma_{12}, \gamma_{13})$ with the reference axes system, the V_2 vector makes an angle $(\gamma_{11}, \gamma_{22}, \gamma_{23})$ with the reference axes and the V_3 vector makes an angle $(\gamma_{31}, \gamma_{32}, \gamma_{33})$ with the reference axes, as shown in Figure 4.9. The set of equations that connects the transformed components of these vectors with the known components of vectors is given by,

$$\overline{V}_1 = V_1 \cos\gamma_{11} + V_2 \cos\gamma_{21} + V_1 \cos\gamma_{31}$$
$$\overline{V}_2 = V_1 \cos\gamma_{12} + V_2 \cos\gamma_{22} + V_1 \cos\gamma_{32} \qquad (4.8)$$
$$\overline{V}_3 = V_1 \cos\gamma_{13} + V_2 \cos\gamma_{23} + V_1 \cos\gamma_{33}$$

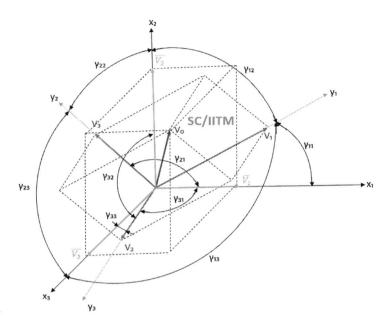

FIGURE 4.9 Components of vector along reference axes.

Let $C_{ij}=\cos \gamma_{ij}$, where, I represents the $(Y_1\text{-}Y_2\text{-}Y_3)$ axes system and j represents the $(X_1\text{-}X_2\text{-}X_3)$ axes system. Thus, the previous equation can be written in matrix form as follows:

$$\begin{Bmatrix} \bar{V}_1 \\ \bar{V}_2 \\ \bar{V}_3 \end{Bmatrix} = \begin{bmatrix} C_{11} & C_{21} & C_{31} \\ C_{12} & C_{22} & C_{32} \\ C_{13} & C_{23} & C_{33} \end{bmatrix} \quad (4.9)$$

$$\{\bar{V}\} = [C_s]^T \{V\} \quad (4.10)$$

where, $[C_s] = \begin{bmatrix} C_{11} & C_{12} & C_{13} \\ C_{21} & C_{22} & C_{23} \\ C_{31} & C_{32} & C_{33} \end{bmatrix}$

It also verifies that $[C_s]^T = [C_s]^{-1}$. $\bar{V}$ refers to the reference axes system and V refers to the coordinate axes system. Please note that they do not refer to the axis of the original vector V_0.

4.5 MEMBER ROTATION MATRIX

Consider a beam element with length L_j. The member has two nodes 'j' and 'k'. The member is oriented along its local axes system $(x_m\text{-}y_m\text{-}z_m)$. This local axes system is placed arbitrarily in space with reference to the standard reference axes system (X-Y-Z).

Three-Dimensional Analysis of Space Frames

γ_x, γ_y, γ_z are the angles of the x_m axis with respect to the X-Y-Z axes respectively, as shown in Figure 4.10.

Let,

$$C_x = \cos \gamma_x$$

$$C_y = \cos \gamma_y$$

$$C_z = \cos \gamma_z$$

As the j and k coordinates of the member positioned in space are known, the direction cosines can be written as follows:

$$C_x = \frac{X_k - X_j}{L_i} \quad (4.11)$$

$$C_y = \frac{Y_k - Y_j}{L_i} \quad (4.12)$$

$$C_z = \frac{Z_k - Z_j}{L_i} \quad (4.13)$$

$$L_i = \sqrt{(X_k - X_j)^2 + (Y_k - Y_j)^2 + (Z_k - Z_j)^2} \quad (4.14)$$

where, (X_k, Y_k, Z_k) and (X_j, Y_j, Z_j) are the coordinates of the beam element placed in space. It is now important to know that the direction cosines give the components of the beam element only along the reference axes system, but an important information

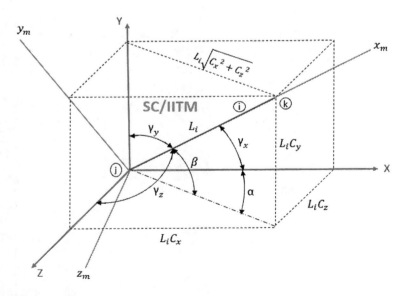

FIGURE 4.10 Member rotation.

of orientation of the local axes system with respect to reference the axes system is not known. Therefore, it is clear that the beam element is oriented along the local axes (x_m-y_m-z_m). Hence, orientation of (x_m-y_m-z_m) axes or the local axes system with respect to the reference axes system is called the ψ angle, which has to be estimated. Thus, in order to find the orientation of the local axes system with respect to the reference axes system, we need to know the direction cosines and ψ angle.

4.6 Y-Z-X TRANSFORMATION

The direction cosines give the components of the member along the reference axes system. Further, the ψ angle provides information about the orientation of the local axes system with respect to the reference axes system. It is very important to note that the procedure of aligning the reference axes system to that of local axes system is called transformation. There are various schemes involved in this transformation. One such scheme is the Y-Z-X transformation, as shown in Figure 4.11, which means rotating the reference axes about the Y-axis, then the Z-axis and finally about the X-axis. Y-Z-X highlights the order or sequence of rotation to be carried out. The reference axes system (X-Y-Z) will be rotated about the Y-axis first, then about the Z-axis and lastly about the X-axis. The amount of rotation happening in the Y and Z axes is α and β respectively, and ultimately we will get the ψ angle after the rotation about the X-axis. The procedure is to hold the orthogonal axes system (X-Y-Z) and rotate this axes system about Y-axis by α degrees.

Let, V_0 be the vector placed arbitrarily. Now, reference axes system is rotated about Y-axis, where Y and Y_α remains the same. The X-axis and Z-axis will move

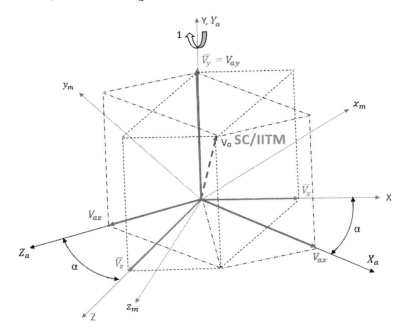

FIGURE 4.11 Y-Z-X transformation.

Three-Dimensional Analysis of Space Frames

to X_α and Z_α respectively, since the angle of rotation is α degrees. $\bar{V}_x, \bar{V}_y, \bar{V}_z$ are the components of the vector with respect to the reference axes (X-Y-Z).

From the figure,

$$\sin \alpha = \frac{C_z}{\sqrt{C_x^2 + C_z^2}}$$

$$\cos \alpha = \frac{C_x}{\sqrt{C_x^2 + C_z^2}}$$

(4.15)

Thus,

$$\begin{Bmatrix} V_{\alpha x} \\ V_{\alpha y} \\ V_{\alpha z} \end{Bmatrix} = \begin{bmatrix} \cos \alpha & 0 & \sin \alpha \\ 0 & 1 & 0 \\ -\sin \alpha & 0 & \cos \alpha \end{bmatrix} \begin{Bmatrix} \bar{V}_x \\ \bar{V}_y \\ \bar{V}_z \end{Bmatrix}$$

By substituting the values of $\cos \alpha$ and $\sin \alpha$ in the previous matrix equation,

$$\begin{Bmatrix} V_{\alpha x} \\ V_{\alpha y} \\ V_{\alpha z} \end{Bmatrix} = \begin{bmatrix} \dfrac{C_x}{\sqrt{C_x^2 + C_z^2}} & 0 & \dfrac{C_z}{\sqrt{C_x^2 + C_z^2}} \\ 0 & 1 & 0 \\ -\dfrac{C_z}{\sqrt{C_x^2 + C_z^2}} & 0 & \dfrac{C_x}{\sqrt{C_x^2 + C_z^2}} \end{bmatrix} \begin{Bmatrix} \bar{V}_x \\ \bar{V}_y \\ \bar{V}_z \end{Bmatrix} \quad (4.16)$$

$$\{V_\alpha\} = [C_\alpha]\{\bar{V}\}$$

The next step is to rotate about the Z-axis by β degrees. Since, the rotation takes place about the Z-axis, Z_α and Z_β remains the same. X_α and Y_α will shift to X_β and Y_β respectively. Now, the vector components are $V_{\beta x}$, $V_{\beta y}$ and $V_{\beta z}$ as shown in Figure 4.12.

Now,

$$\begin{Bmatrix} V_{\beta x} \\ V_{\beta y} \\ V_{\beta z} \end{Bmatrix} = \begin{bmatrix} \cos \beta & \sin \beta & 0 \\ -\sin \beta & \cos \beta & 0 \\ 0 & 0 & 1 \end{bmatrix} \begin{Bmatrix} V_{\alpha x} \\ V_{\alpha y} \\ V_{\alpha z} \end{Bmatrix}$$

$$\sin \beta = C_y$$

$$\cos \beta = \sqrt{C_x^2 + C_z^2}$$

By substituting these values,

$$\begin{Bmatrix} V_{\beta x} \\ V_{\beta y} \\ V_{\beta z} \end{Bmatrix} = \begin{bmatrix} \sqrt{C_x^2 + C_z^2} & C_y & 0 \\ -C_y & \sqrt{C_x^2 + C_z^2} & 0 \\ 0 & 0 & 1 \end{bmatrix} \begin{Bmatrix} V_{\alpha x} \\ V_{\alpha y} \\ V_{\alpha z} \end{Bmatrix}$$

$$\{V_\beta\} = [C_\beta]\{V_\alpha\} \quad (4.17)$$

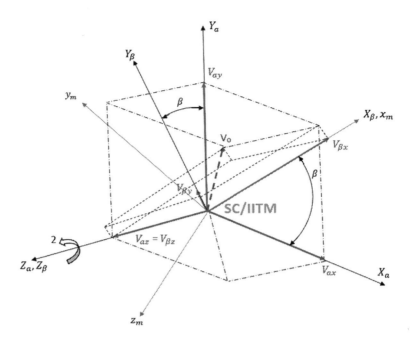

FIGURE 4.12 Z-axis rotation.

Finally, let us rotate about the X-axis by an angle ψ_y as shown in Figure 4.13. Here X_β will remains same as x_m. Y_β and Z_β will shift to y_m and z_m respectively. Thus, ψ_y is the angle between Y_β and y_m measured from Y_β toward y_m or angle between Z_β and z_m measured Z_β toward z_m. The idea is to bring y_m and Y_β aligned and x_m and X_β aligned. x_m and X_β are already aligned, since the rotation is about the X axis.

Thus,

$$\begin{Bmatrix} V_x \\ V_y \\ V_z \end{Bmatrix} = \begin{bmatrix} 1 & 0 & 0 \\ 0 & \cos\psi_y & \sin\psi_y \\ 0 & -\sin\psi_y & \cos\psi_y \end{bmatrix} \begin{Bmatrix} V_{\beta x} \\ V_{\beta y} \\ V_{\beta z} \end{Bmatrix}$$

$$\{V\} = \begin{bmatrix} C_{\psi_y} \end{bmatrix}\{V_\beta\}$$

Hence,

$$\{V\} = \begin{bmatrix} C_{\psi_y} \end{bmatrix}\{V_\beta\} = \begin{bmatrix} C_{\psi_y} \end{bmatrix}\begin{bmatrix} C_\beta \end{bmatrix}\{V_\alpha\} = \begin{bmatrix} C_{\psi_y} \end{bmatrix}\begin{bmatrix} C_\beta \end{bmatrix}\begin{bmatrix} C_\alpha \end{bmatrix}\{\bar{V}\} \quad (4.18)$$

$$\{V\} = \begin{bmatrix} C_{\psi_y} \end{bmatrix}\begin{bmatrix} C_\beta \end{bmatrix}\begin{bmatrix} C_\alpha \end{bmatrix}\{\bar{V}\} \quad (4.19)$$

$$\text{Let,} \begin{bmatrix} C_y \end{bmatrix} = \begin{bmatrix} C_{\psi_y} \end{bmatrix}\begin{bmatrix} C_\beta \end{bmatrix}\begin{bmatrix} C_\alpha \end{bmatrix}$$

Three-Dimensional Analysis of Space Frames

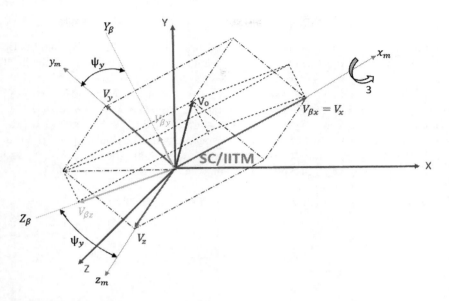

FIGURE 4.13 Rotation about X-axis.

By substituting the values,

$$[C_y] = \begin{bmatrix} 1 & 0 & 0 \\ 0 & \cos\psi_y & \sin\psi_y \\ 0 & -\sin\psi_y & \cos\psi_y \end{bmatrix} \begin{bmatrix} \sqrt{C_x^2 + C_z^2} & C_y & 0 \\ -C_y & \sqrt{C_x^2 + C_z^2} & 0 \\ 0 & 0 & 1 \end{bmatrix}$$

$$\begin{bmatrix} \dfrac{C_x}{\sqrt{C_x^2 + C_z^2}} & 0 & \dfrac{C_z}{\sqrt{C_x^2 + C_z^2}} \\ 0 & 1 & 0 \\ -\dfrac{C_z}{\sqrt{C_x^2 + C_z^2}} & 0 & \dfrac{C_x}{\sqrt{C_x^2 + C_z^2}} \end{bmatrix}$$

$$[C_y] = \begin{bmatrix} C_x & C_y & C_z \\ \dfrac{-C_x C_y \cos\psi_y - C_z \sin\psi_y}{\sqrt{C_x^2 + C_z^2}} & \cos\psi_y \sqrt{C_x^2 + C_z^2} & \dfrac{-C_z C_y \cos\psi_y + C_x \sin\psi_y}{\sqrt{C_x^2 + C_z^2}} \\ \dfrac{C_x C_y \sin\psi_y - C_z \sin\psi_y}{\sqrt{C_x^2 + C_y^2}} & -\sin\psi_y \sqrt{C_x^2 + C_z^2} & \dfrac{C_y C_z \sin\psi_y - C_x \cos\psi_y}{\sqrt{C_x^2 + C_z^2}} \end{bmatrix}$$

$[C_y]$ is called as the rotation matrix, which is expressed in terms if direction cosines. It also defines the position of x_m or the longitudinal axis of the member in the local axes system with θ, in relation to the reference axes system and through the ψ angle.

4.7 Z-Y-X TRANSFORMATION

In this transformation, rotate about the Z-axis first, then about the Y-axis and finally about the X-axis. Here, one can get the ψ_z angle. As shown in Figure 4.14, the local axes system is (x_m, y_m, z_m) and the global axes system is (X-Y-Z).

When rotated about the Z-axis by an angle of θ degrees, the Z and Z_θ axes will remain same. X and Y will shift to X_θ and Y_θ respectively. In the second step, rotation about the Y-axis by an angle of α degrees, will leave Y_θ and Y_λ axes the same. X_θ and Z_θ will shift to X_λ and Z_λ respectively. In the final step, by rotating about the X-axis by an angle of ψ_z, X_λ and x_m will remain the same. But, Y_λ and Z_λ will shift to y_m and z_m respectively. Thus, ψ_z is measured from Y_λ to y_m or Z_λ to z_m.

Hence, the rotation matrix can be directly written as follows:

$$[C_{\psi_z}] = \begin{bmatrix} 1 & 0 & 0 \\ 0 & \cos\psi_z & \sin\psi_z \\ 0 & -\sin\psi_z & \cos\psi_z \end{bmatrix}$$

We can also write,

$$[C_z] = [C_{\psi_z}][C_\lambda][C_\theta] \quad (4.20)$$

$$[C_z] = \begin{bmatrix} \cos\theta\cos\lambda & \sin\theta\cos\lambda & \sin\lambda \\ (-\sin\theta\cos\psi_z)-(\cos\theta\sin\lambda\sin\psi_z) & (\cos\theta\cos\psi_z)-(\sin\theta\sin\lambda\sin\psi_z) & \cos\lambda\sin\psi_z \\ (\sin\theta\sin\psi_z)-(\cos\theta\sin\lambda\cos\psi_z) & (-\cos\theta\sin\psi_z)-(\sin\theta\sin\lambda\cos\psi_z) & \cos\lambda\cos\psi_z \end{bmatrix}$$

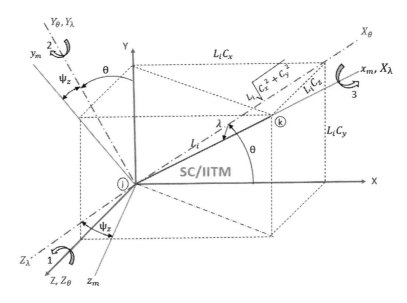

FIGURE 4.14 X-Y-Z transformation.

From the figure,

$$\sin\theta = \frac{C_y}{\sqrt{C_x^2 + C_y^2}}$$

$$\cos\theta = \frac{C_x}{\sqrt{C_x^2 + C_y^2}}$$

$$\sin\lambda = C_z$$

$$\cos\lambda = \sqrt{C_x^2 + C_y^2}$$

By substituting the previously mentioned values in C_z,

$$[C_z] = \begin{bmatrix} C_x & C_y & C_z \\ \dfrac{(-C_x C_z \sin\psi_z) - (C_y \cos\psi_z)}{\sqrt{C_x^2 + C_y^2}} & \dfrac{(-C_y C_z \sin\psi_z) - (C_x \cos\psi_z)}{\sqrt{C_x^2 + C_y^2}} & \dfrac{\sin\psi_z}{\sqrt{C_x^2 + C_y^2}} \\ \dfrac{(-C_x C_z \cos\psi_z) - (C_y \sin\psi_z)}{\sqrt{C_x^2 + C_y^2}} & \dfrac{(-C_y C_z \cos\psi_z) - (C_x \sin\psi_z)}{\sqrt{C_x^2 + C_y^2}} & \dfrac{\cos\psi_z}{\sqrt{C_x^2 + C_y^2}} \end{bmatrix}$$

In general, one can use either Y-Z-X transformation or Z-Y-X transformation to transform the reference axes system to the local axes system. But, choice of transformation order can make the evaluation of the ψ angle, simpler. In both the transformations, the rotation in the last step takes place about the X-axis only. Most importantly, in both the transformations, ψ angles are calculated. But, ψ_y and ψ_z are completely different. Thus, for a given member, which transformation order is to be followed? If the member is positioned in the frame of reference axes, such that the longitudinal axes of the member corresponds to Y-axis of the reference system, then use the Z-Y-X transformation. Similarly, if a member is placed in the frame of the reference axes system, such that the longitudinal axes of the member corresponds to Z-axis of the reference system, then use Y-Z-X transformation.

4.8 THE Ψ ANGLE

Let us consider a three-axes system (X-Y-Z), as shown in Figure 4.15. The local axes system is (x_m-y_m-z_m). The transformed axes system is (X_β, Y_β, Z_β). The line of sight is along the X-axis. The longitudinal axis of the member coincides with the X-axis. Thus, this is the Y-Z-X transformation. The axes are then marked on the cross-section of the member seen from the line of sight. The angle ψ_y is measured anticlockwise when viewing the cross-section of the member toward the *j*th end from the *k*th end. So, the direction cosines define the location of the x_m axis. ψ_y defines the location of the minor principal axis. All parameters are geometric dependent.

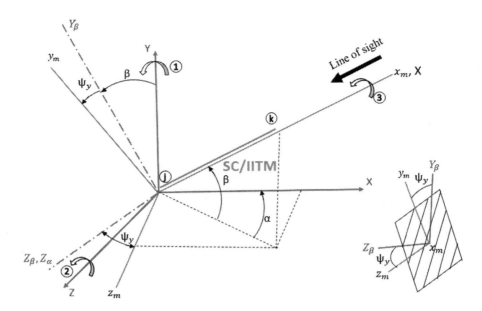

FIGURE 4.15 Y-Z-X transformation.

Similarly, for the Z-Y-X transformation, as shown in Figure 4.16, the line of sight is along the x_m axis. The longitudinal axis of the member coincides with the x_m axis. Here, ψ_z is measured anticlockwise when viewing the cross-section of the member toward the jth end from the kth end.

4.9 ANALYSIS OF SPACE FRAME

For any member, which is arbitrarily oriented in space with respect to the reference axes system, one needs to estimate the two parameters:

1. Direction cosines
2. ψ angle

The important points to be noted in the analysis of space frame are as follows:

1. Three-direction cosines define the location of the longitudinal axis of the member (x_m) axis with respect to the reference axes system.
2. The ψ angle defines the location of the minor principal axis.
3. All parameters of direction cosines and ψ angle are geometric dependent. It actually depends on the position or orientation of the member with respect to the reference axes system.
4. Direction cosines of each member can be readily computed, but the ψ angle need to be carefully estimated.

Three-Dimensional Analysis of Space Frames

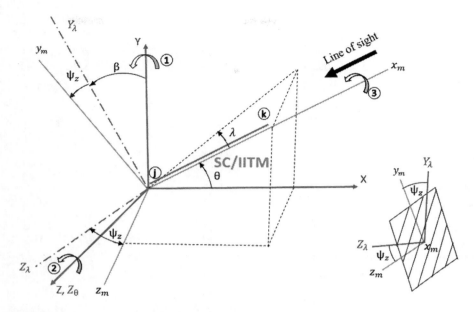

FIGURE 4.16 Z-Y-X transformation.

5. One can do two types of transformations or rotations: Y-Z-X and Z-Y-X transformation. The ψ angle computed from both the transformations will be different. Obviously, the rotation process enables alignment of the X-Y-Z axes to x-y-z axes of the member.

The primary objective is to extend the knowledge of two-dimensional analysis to three-dimensional analysis. Consider a fixed beam element arbitrarily oriented with two nodes 'j' and 'k'. The member is designated as 'i', as shown in Figure 4.17. Each node will have three translations and three rotational degrees-of-freedom. The beam

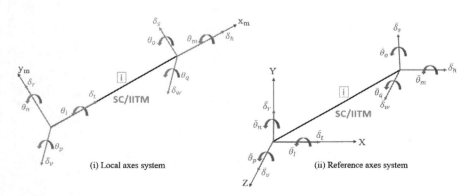

FIGURE 4.17 Local and reference axes system.

element will have twelve degrees-of-freedom, making the stiffness matrix of size 12×12, unlike two-dimensional analysis.

It is assumed that the orientation of the local axes system with respect to the reference axes system can be described in terms of the direction cosines C_x, C_y, C_z and the ψ angle. We already know that,

$$C_x = \frac{X_k - X_j}{L_i}$$

$$C_y = \frac{Y_k - Y_j}{L_i}$$

$$C_z = \frac{Z_k - Z_j}{L_i}$$

$$L_i = \sqrt{(X_k - X_j)^2 + (Y_k - Y_j)^2 + (Z_k - Z_j)^2}$$

The set of translations at the *j*th end measured in the local axes system can be connected to the reference axes system as follows:

$$\begin{Bmatrix} \delta_t \\ \delta_r \\ \delta_r \end{Bmatrix}_i = \begin{bmatrix} C_{11} & C_{12} & C_{13} \\ C_{21} & C_{22} & C_{23} \\ C_{31} & C_{32} & C_{33} \end{bmatrix} \begin{Bmatrix} \overline{\delta}_t \\ \overline{\delta}_r \\ \overline{\delta}_r \end{Bmatrix}_i$$

It can be seen that the previous equation gives the translation in the *j*th end. Similarly, the set of rotations measured at the *j*th end can be translated from the local axes to the reference axes system as follows:

$$\begin{Bmatrix} \theta_l \\ \theta_n \\ \theta_p \end{Bmatrix}_i = \begin{bmatrix} C_{11} & C_{12} & C_{13} \\ C_{21} & C_{22} & C_{23} \\ C_{31} & C_{32} & C_{33} \end{bmatrix} \begin{Bmatrix} \overline{\theta}_l \\ \overline{\theta}_n \\ \overline{\theta}_p \end{Bmatrix}_i$$

Similarly, for the *k*th end,

$$\begin{Bmatrix} \delta_h \\ \delta_s \\ \delta_w \end{Bmatrix}_i = \begin{bmatrix} C_{11} & C_{12} & C_{13} \\ C_{21} & C_{22} & C_{23} \\ C_{31} & C_{32} & C_{33} \end{bmatrix} \begin{Bmatrix} \overline{\delta}_h \\ \overline{\delta}_s \\ \overline{\delta}_w \end{Bmatrix}_i$$

$$\begin{Bmatrix} \theta_m \\ \theta_o \\ \theta_q \end{Bmatrix}_i = \begin{bmatrix} C_{11} & C_{12} & C_{13} \\ C_{21} & C_{22} & C_{23} \\ C_{31} & C_{32} & C_{33} \end{bmatrix} \begin{Bmatrix} \overline{\theta}_m \\ \overline{\theta}_o \\ \overline{\theta}_q \end{Bmatrix}_i$$

Three-Dimensional Analysis of Space Frames

By combining all twelve degrees-of-freedom,

$$\begin{Bmatrix} \delta_t \\ \delta_r \\ \delta_v \\ \theta_l \\ \theta_n \\ \theta_p \\ \delta_h \\ \delta_s \\ \delta_w \\ \theta_m \\ \theta_o \\ \theta_q \end{Bmatrix} = \begin{bmatrix} C_{11} & C_{12} & C_{13} & 0 & 0 & 0 & 0 & 0 & 0 & 0 & 0 & 0 \\ C_{21} & C_{22} & C_{23} & 0 & 0 & 0 & 0 & 0 & 0 & 0 & 0 & 0 \\ C_{31} & C_{32} & C_{33} & 0 & 0 & 0 & 0 & 0 & 0 & 0 & 0 & 0 \\ 0 & 0 & 0 & C_{11} & C_{12} & C_{13} & 0 & 0 & 0 & 0 & 0 & 0 \\ 0 & 0 & 0 & C_{21} & C_{22} & C_{23} & 0 & 0 & 0 & 0 & 0 & 0 \\ 0 & 0 & 0 & C_{31} & C_{32} & C_{33} & 0 & 0 & 0 & 0 & 0 & 0 \\ 0 & 0 & 0 & 0 & 0 & 0 & C_{11} & C_{12} & C_{13} & 0 & 0 & 0 \\ 0 & 0 & 0 & 0 & 0 & 0 & C_{21} & C_{22} & C_{23} & 0 & 0 & 0 \\ 0 & 0 & 0 & 0 & 0 & 0 & C_{31} & C_{32} & C_{33} & 0 & 0 & 0 \\ 0 & 0 & 0 & 0 & 0 & 0 & 0 & 0 & 0 & C_{11} & C_{12} & C_{13} \\ 0 & 0 & 0 & 0 & 0 & 0 & 0 & 0 & 0 & C_{21} & C_{22} & C_{23} \\ 0 & 0 & 0 & 0 & 0 & 0 & 0 & 0 & 0 & C_{31} & C_{32} & C_{33} \end{bmatrix} \begin{Bmatrix} \bar{\delta}_t \\ \bar{\delta}_r \\ \bar{\delta}_v \\ \bar{\theta}_l \\ \bar{\theta}_n \\ \bar{\theta}_p \\ \bar{\delta}_h \\ \bar{\delta}_s \\ \bar{\delta}_w \\ \bar{\theta}_m \\ \bar{\theta}_o \\ \bar{\theta}_q \end{Bmatrix}$$

Thus, $\{\delta^{(s)}\}_i = [T_i^{(s)}]\{\bar{\delta}^{(s)}\}_i$ (4.21)

where,

$$[T_i] = \begin{bmatrix} [C_i] & [0] & [0] & [0] \\ [0] & [C_i] & [0] & [0] \\ [0] & [0] & [C_i] & [0] \\ [0] & [0] & [0] & [C_i] \end{bmatrix}_{12 \times 12}$$

This is the transformation matrix in three-dimensional space. This transformation matrix also has some special properties similar to those of the transformation matrix in two-dimensional space.

$$[T^{(s)}]_i^T = [T^{(s)}]_i^{-1} \qquad (4.22)$$

Hence, $[C_i]^T = [C_i]^{-1}$ (4.23)

Therefore,

$$[\bar{\delta}_i] = [T]_i^T \{\delta\}_i \qquad (4.24)$$

Where, [C] is the rotation matrix, whose elements are the direction cosines. Thus, C_y can be used for Y-Z-X transformation and C_z for Z-Y-X transformation, which contains the direction cosines and ψ angle.

In addition, the following equations used in two-dimensional analysis are also valid for three-dimensional analysis.

$$\{m\}_i = [T]_i \{\bar{m}\}_i$$

$$[\bar{K}]_i = [T^{(s)}]_i^T [K^{(s)}]_i [T^{(s)}]_i \qquad (4.25)$$

$$[K_{uu}]\{\Delta_u\} = \{J_L\}_u$$

$$\{M\}_i = [K]_i [T]\{\bar{\delta}\} + \{FEM\}_i$$

Example problems with computer program

EXAMPLE 4.1:

Determine the ψ angle for the members of space frame shown in Figure 4.18.

SOLUTION:

Mark the global axis and local axes for all the members.

Member	j	k	Length (m)
1	1	2	4
2	3	2	4
3	2	4	6.402

Member 1:

The x_m axis of member 1 is toward the Y-axis. Therefore, we should use Z-Y-X transformation.

Here, ψ angle = ψ_z = angle between y_m and Y-axis = 90 degrees

Member 2:

The x_m axis of member 2 is toward the X-axis. Therefore, we can use either Y-Z-X or Z-Y-X transformation.

Let us use Y-Z-X transformation.

Here, ψ angle = ψ_y = angle between y_m and Y-axis = 90 degrees

Member 3:

The x_m axis of member 2 is not aligned toward X, Y and Z but it is inclined. Let us try the Y-Z-X transformation.

To understand this transformation, let us consider a simple example. Consider a vector that is placed arbitrarily in the space, as shown in Figure 4.19. Let

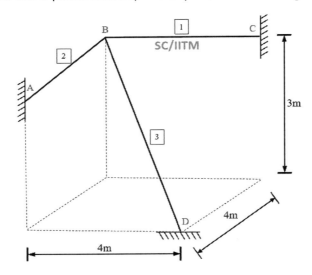

FIGURE 4.18 Space frame example.

Three-Dimensional Analysis of Space Frames

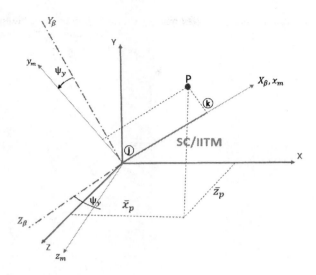

FIGURE 4.19 Transformation.

us consider a point 'p' in the x_m-y_m plane in the frame of X-Y-Z, but the point should not be located along the x_m axis. The point is projected on other planes. The coordinates of this point will describe the position of $\bar{X}_p$, $\bar{Y}_p$, $\bar{Z}_p$ with respect to the reference axes.

For the Y-Z-X transformation, we already know that,

$$\begin{Bmatrix} X_{\beta p} \\ Y_{\beta p} \\ Z_{\beta p} \end{Bmatrix} = \begin{bmatrix} \sqrt{C_x^2 + C_z^2} & C_y & 0 \\ -C_y & \sqrt{C_x^2 + C_z^2} & 0 \\ 0 & 0 & 1 \end{bmatrix} \begin{bmatrix} \dfrac{C_x}{\sqrt{C_x^2 + C_z^2}} & 0 & \dfrac{C_z}{\sqrt{C_x^2 + C_z^2}} \\ 0 & 1 & 0 \\ -\dfrac{C_z}{\sqrt{C_x^2 + C_z^2}} & 0 & \dfrac{C_x}{\sqrt{C_x^2 + C_z^2}} \end{bmatrix} \begin{Bmatrix} \bar{X}_p \\ \bar{Y}_p \\ \bar{Z}_p \end{Bmatrix}$$

$$\begin{Bmatrix} X_{\beta p} \\ Y_{\beta p} \\ Z_{\beta p} \end{Bmatrix} = \begin{bmatrix} C_x & C_y & C_z \\ \dfrac{-C_x C_y}{\sqrt{C_x^2 + C_z^2}} & \sqrt{C_x^2 + C_z^2} & \dfrac{-C_y C_z}{\sqrt{C_x^2 + C_z^2}} \\ -\dfrac{C_z}{\sqrt{C_x^2 + C_z^2}} & 0 & \dfrac{C_x}{\sqrt{C_x^2 + C_z^2}} \end{bmatrix} \begin{Bmatrix} \bar{X}_p \\ \bar{Y}_p \\ \bar{Z}_p \end{Bmatrix}$$

The coordinates of the point 'p' on the (x_m, y_m) plane can be written as $Y_{\beta p}$, $Z_{\beta p}$. Thus,

$$\sin \psi_y = \frac{Z_{\beta p}}{\sqrt{Z_{\beta p}^2 + Y_{\beta p}^2}}$$

$$\cos \psi_y = \frac{Y_{\beta p}}{\sqrt{Z_{\beta p}^2 + Y_{\beta p}^2}}$$

So, the coordinates of point 'p' with respect to the reference axes system should be P(4,4,0). Similarly, the coordinates of point 'p' with respect to the *j*th end of the member of 3 is P(0,0, −3). Thus, the position vector is given by,

$$\begin{Bmatrix} \overline{X}_p \\ \overline{Y}_p \\ \overline{Z}_p \end{Bmatrix} = \begin{Bmatrix} 0 \\ 0 \\ -3 \end{Bmatrix}$$

Now, the coordinates of the *j*th end with respect to X-Y-Z axes system are (4,4,3). Similarly, the coordinates of the *k*th end with respect to the X-Y-Z system are (0,0,0).

1. *Direction cosines:*

$$C_x = \frac{X_k - X_j}{L_i} = \frac{0-4}{6.402} = -0.625$$

$$C_y = \frac{Y_k - Y_j}{L_i} = \frac{0-4}{6.402} = -0.625$$

$$C_z = \frac{Z_k - Z_j}{L_i} = \frac{0-3}{6.402} = -0.469$$

2. ψ *angle:*
Now,

$$X_{\beta p} = C_x \overline{X}_p + C_y \overline{Y}_p + C_z \overline{Z}_p = 0 + 0 + (0.469 \times 3) = 1.407$$

$$Y_{\beta p} = \frac{-C_x C_y}{\sqrt{C_x^2 + C_z^2}} \overline{X}_p + \sqrt{C_x^2 + C_z^2}\, \overline{Y}_p - \frac{C_y C_z}{\sqrt{C_x^2 + C_z^2}} \overline{Z}_p = 0 + 0 - \frac{-(-0.625)(-0.469)(-3)}{\sqrt{0.625^2 + 0.429^2}} = +1.125$$

$$Z_{\beta p} = -\frac{C_z}{\sqrt{C_x^2 + C_z^2}} \overline{X}_p + \frac{C_x}{\sqrt{C_x^2 + C_z^2}} \overline{Z}_p = 0 + \frac{(-0.625)(-3)}{\sqrt{0.625^2 + 0.429^2}} = +2.40$$

$$\sin \psi_y = \frac{Z_{\beta p}}{\sqrt{Z_{\beta p}^2 + Y_{\beta p}^2}} = 0.905$$

$$\cos \psi_y = \frac{Y_{\beta p}}{\sqrt{Z_{\beta p}^2 + Y_{\beta p}^2}} = 0.424$$

Thus,

$$\psi_y = \tan^{-1}\left\{\frac{0.905}{0.424}\right\} = 64.897° + 180° = 244.897°$$

EXAMPLE 4.2:

Determine the ψ angle for the members of the space frame shown in Figure 4.20.

SOLUTION:

Mark the global axis and local axes for all the members.

Three-Dimensional Analysis of Space Frames

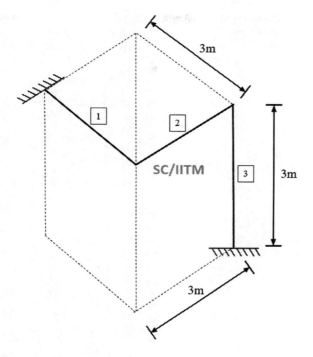

FIGURE 4.20 Space frame example.

1. *Coordinates of joints:*

Joint	X	Y	Z
A	0	0	0
B	3	0	0
C	3	0	−3
D	3	−3	−3

2. ψ *angle:*

Member	Length (m)	Joint j	Joint k	Direction Cosines C_x	C_y	C_z	Type of Transformation	ψ Angle (degrees)
AB	3	A	B	1	0	0	Y-Z-X	$\psi_y = 0$
BC	3	B	C	0	0	−1	Y-Z-X	$\psi_y = 0$
CD	3	D	C	0	1	0	Z-Y-X	$\psi_z = 90$

Member 1: Inclination angles of x_m with X, Y, Z are (0,90,90) degrees.
Member 2: Inclination angles of x_m with X, Y, Z are (90,90,0) degrees.
Member 3: Inclination angles of x_m with X, Y, Z are (90,0,90) degrees.

EXAMPLE 4.3:

Analyze the three-dimensional space frame, as shown in Figure 4.21, using the stiffness method of analysis.

SOLUTION:

Mark the local and reference axes system.

1. *Coordinates of joints:*

Joint	X	Y	Z
A	0	0	0
B	3	0	0
C	3	0	−3
D	3	−3	−3

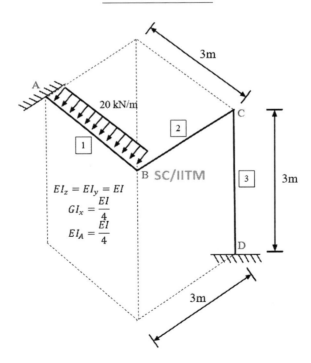

FIGURE 4.21 Space frame example.

Three-Dimensional Analysis of Space Frames

2. ψ angle:

Member	Length (m)	Joint j	Joint k	C_x	C_y	C_z	Type of Transformation	ψ Angle (degrees)
AB	3	A	B	1	0	0	Y-Z-X	$\psi_y = 0$
BC	3	B	C	0	0	−1	Y-Z-X	$\psi_y = 0$
CD	3	D	C	0	1	0	Z-Y-X	$\psi_z = 90$

3 *Marking unrestrained and restrained degrees-of-freedom:*
The unrestrained and restrained degrees-of-freedom are shown in Figure 4.22.
Unrestrained degrees-of-freedom: 12 ($\delta_1, \delta_2, \delta_3, \theta_4, \theta_5, \theta_6, \delta_7, \delta_8, \delta_9, \theta_{10}, \theta_{11}, \theta_{12}$)
Restrained degrees-of-freedom: 12 ($\bar{\delta}_{13}, \bar{\delta}_{14}, \bar{\delta}_{15}, \bar{\theta}_{16}, \bar{\theta}_{17}, \bar{\theta}_{18}, \bar{\delta}_{19}, \bar{\delta}_{20}, \bar{\delta}_{21}, \bar{\theta}_{22}, \bar{\theta}_{23}, \bar{\theta}_{24}$)

4. *Estimation of transformation matrix:*
We already know that,

$$C = \begin{bmatrix} C_{11} & C_{12} & C_{13} \\ C_{21} & C_{22} & C_{23} \\ C_{31} & C_{32} & C_{33} \end{bmatrix}$$

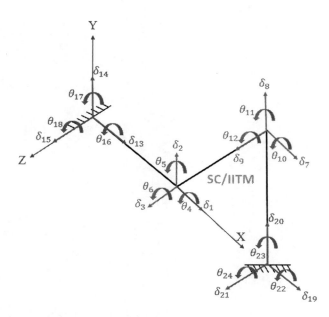

FIGURE 4.22 Unrestrained and restrained degrees-of-freedom.

The direction cosine matrices for all the three members are,

$$C_1 = \begin{bmatrix} 1 & 0 & 0 \\ 0 & 1 & 0 \\ 0 & 0 & 1 \end{bmatrix}, \quad C_2 = \begin{bmatrix} 0 & 0 & -1 \\ 0 & 1 & 0 \\ 1 & 0 & 0 \end{bmatrix}, \quad C_3 = \begin{bmatrix} 0 & 1 & 0 \\ 0 & 0 & 1 \\ 1 & 0 & 0 \end{bmatrix}$$

Now, the transformation matrix is given by,

$$[T_i] = \begin{bmatrix} [C_i] & [0] & [0] & [0] \\ [0] & [C_i] & [0] & [0] \\ [0] & [0] & [C_i] & [0] \\ [0] & [0] & [0] & [C_i] \end{bmatrix}_{12 \times 12}$$

Global labels:

AB = [13, 14, 15, 16, 17, 18, 1, 2, 3, 4, 5, 6]
BC = [1, 2, 3, 4, 5, 6, 7, 8, 9, 10, 11, 12]
CD = [19, 20, 21, 22, 23, 24, 7, 8, 9, 10, 11, 12]

5. *Fixed end moments and joint load vector:*
   Load is acting only on the member AB. Hence, the fixed end moments along the other two members remains zero.

$$(FEM)_{AB} = \begin{Bmatrix} 0 \\ 30 \\ 0 \\ 0 \\ 0 \\ 15 \\ 0 \\ 30 \\ 0 \\ 0 \\ 0 \\ -15 \end{Bmatrix}$$

The joint load vector is the reversal of the fixed end moments. The transpose of the joint load vector is given as follows:

$$[J_L]^T = \{0 \ -30 \ 0 \ 0 \ 0 \ 15 \ 0 \ 0 \ 0 \ 0 \ 0 \ 0 \ -30 \ 0 \ 0 \ 0 \ -15 \ 0 \ 0 \ 0 \ 0 \ 0 \ 0\}$$

Three-Dimensional Analysis of Space Frames

6. *Stiffness matrix:*

$$[K_{AB}] = EI \begin{bmatrix} 0.083 & 0 & 0 & 0 & 0 & 0 & -0.083 & 0 & 0 & 0 & 0 & 0 \\ 0 & 0.444 & 0 & 0 & 0 & 0.667 & 0 & -0.444 & 0 & 0 & 0 & 0.667 \\ 0 & 0 & 0.444 & 0 & -0.667 & 0 & 0 & 0 & 0.444 & 0 & 0.667 & 0 \\ 0 & 0 & 0 & 0.083 & 0 & 0 & 0 & 0 & 0 & -0.083 & 0 & 0 \\ 0 & 0 & -0.667 & 0 & 1.333 & 0 & 0 & 0 & 0.667 & 0 & 0.667 & 0 \\ 0 & 0.667 & 0 & 0 & 0 & 1.333 & 0 & -0.667 & 0 & 0 & 0 & 0.667 \\ -0.083 & 0 & 0 & 0 & 0 & 0 & -0.083 & 0 & 0 & 0 & 0 & 0 \\ 0 & -0.444 & 0 & 0 & 0 & -0.667 & 0 & 0.444 & 0 & 0 & 0 & -0.667 \\ 0 & 0 & -0.444 & 0 & 0.667 & 0 & 0 & 0 & 0.444 & 0 & 0.667 & 0 \\ 0 & 0 & 0 & -0.083 & 0 & 0 & 0 & 0 & 0 & 0.083 & 0 & 0 \\ 0 & 0 & -0.667 & 0 & 0.667 & 0 & 0 & 0 & 0.667 & 0 & 1.333 & 0 \\ 0 & 0.667 & 0 & 0 & 0 & 0.667 & 0 & -0.667 & 0 & 0 & 0 & 1.333 \end{bmatrix}$$

$$[\bar{K}_{AB}] = EI \begin{bmatrix} 0.083 & 0 & 0 & 0 & 0 & 0 & -0.083 & 0 & 0 & 0 & 0 & 0 \\ 0 & 0.444 & 0 & 0 & 0 & 0.667 & 0 & -0.444 & 0 & 0 & 0 & 0.667 \\ 0 & 0 & 0.444 & 0 & -0.667 & 0 & 0 & 0 & 0.444 & 0 & 0.667 & 0 \\ 0 & 0 & 0 & 0.083 & 0 & 0 & 0 & 0 & 0 & -0.083 & 0 & 0 \\ 0 & 0 & -0.667 & 0 & 1.333 & 0 & 0 & 0 & 0.667 & 0 & 0.667 & 0 \\ 0 & 0.667 & 0 & 0 & 0 & 1.333 & 0 & -0.667 & 0 & 0 & 0 & 0.667 \\ -0.083 & 0 & 0 & 0 & 0 & 0 & -0.083 & 0 & 0 & 0 & 0 & 0 \\ 0 & -0.444 & 0 & 0 & 0 & -0.667 & 0 & 0.444 & 0 & 0 & 0 & -0.667 \\ 0 & 0 & -0.444 & 0 & 0.667 & 0 & 0 & 0 & 0.444 & 0 & 0.667 & 0 \\ 0 & 0 & 0 & -0.083 & 0 & 0 & 0 & 0 & 0 & 0.083 & 0 & 0 \\ 0 & 0 & -0.667 & 0 & 0.667 & 0 & 0 & 0 & 0.667 & 0 & 1.333 & 0 \\ 0 & 0.667 & 0 & 0 & 0 & 0.667 & 0 & -0.667 & 0 & 0 & 0 & 1.333 \end{bmatrix}$$

$$[K_{BC}] = [K_{AB}]$$

$$[\bar{K}_{BC}] = EI \begin{bmatrix} 0.444 & 0 & 0 & 0 & -0.667 & 0 & -0.444 & 0 & 0 & 0 & -0.667 & 0 \\ 0 & 0.444 & 0 & 0.667 & 0 & 0 & 0 & -0.444 & 0 & 0.667 & 0 & 0 \\ 0 & 0 & 0.083 & 0 & 0 & 0 & 0 & 0 & -0.083 & 0 & 0 & 0 \\ 0 & 0.667 & 0 & 1.333 & 0 & 0 & 0 & -0.667 & 0 & 0.667 & 0 & 0 \\ -0.667 & 0 & 0 & 0 & 1.333 & 0 & 0.667 & 0 & 0 & 0 & 0.667 & 0 \\ 0 & 0 & 0 & 0 & 0 & 0.083 & 0 & 0 & 0 & 0 & 0 & -0.083 \\ -0.444 & 0 & 0 & 0 & 0.667 & 0 & 0.444 & 0 & 0 & 0 & 0.667 & 0 \\ 0 & -0.444 & 0 & -0.667 & 0 & 0 & 0 & 0.444 & 0 & -0.667 & 0 & 0 \\ 0 & 0 & -0.083 & 0 & 0 & 0 & 0 & 0 & 0.083 & 0 & 0 & 0 \\ 0 & 0.667 & 0 & 0.667 & 0 & 0 & 0 & -0.667 & 0 & 1.333 & 0 & 0 \\ -0.667 & 0 & 0 & 0 & 0.667 & 0 & 0.667 & 0 & 0 & 0 & 1.333 & 0 \\ 0 & 0 & 0 & 0 & 0 & -0.083 & 0 & 0 & 0 & 0 & 0 & 0.083 \end{bmatrix}$$

$$[K_{CD}] = [K_{AB}]$$

$$[\bar{K}_{CD}] = EI \begin{bmatrix} 0.444 & 0 & 0 & 0 & 0 & -0.667 & -0.444 & 0 & 0 & 0 & 0 & -0.667 \\ 0 & 0.083 & 0 & 0 & 0 & 0 & 0 & -0.083 & 0 & 0 & 0 & 0 \\ 0 & 0 & 0.444 & 0.667 & 0 & 0 & 0 & 0 & -0.444 & 0.667 & 0 & 0 \\ 0 & 0 & 0.667 & 1.333 & 0 & 0 & 0 & 0 & -0.667 & 0.667 & 0 & 0 \\ 0 & 0 & 0 & 0 & 0.083 & 0 & 0 & 0 & 0 & 0 & -0.083 & 0 \\ -0.667 & 0 & 0 & 0 & 0 & 1.333 & 0.667 & 0 & 0 & 0 & 0 & 0.667 \\ -0.444 & 0 & 0 & 0 & 0 & 0.667 & 0.444 & 0 & 0 & 0 & 0 & 0.667 \\ 0 & -0.083 & 0 & 0 & 0 & 0 & 0 & 0.083 & 0 & 0 & 0 & 0 \\ 0 & 0 & -0.444 & -0.667 & 0 & 0 & 0 & 0 & 0.444 & -0.667 & 0 & 0 \\ 0 & 0 & 0.667 & 0.667 & 0 & 0 & 0 & 0 & -0.667 & 1.333 & 0 & 0 \\ 0 & 0 & 0 & 0 & -0.083 & 0 & 0 & 0 & 0 & 0 & 0.083 & 0 \\ -0.667 & 0 & 0 & 0 & 0 & 0.667 & 0.667 & 0 & 0 & 0 & 0 & 1.333 \end{bmatrix}$$

The total stiffness matrix can be formed by assembling the global stiffness matrices of all the members, from which the unrestrained stiffness matrix can be partitioned.

$$[K_{uu}] = EI \begin{bmatrix} 0.528 & 0 & 0 & 0 & -0.667 & 0 & -0.444 & 0 & 0 & 0 & -0.667 & 0 \\ 0 & 0.889 & 0 & 0.667 & 0 & -0.667 & 0 & -0.444 & 0 & 0.667 & 0 & 0 \\ 0 & 0 & 0.528 & 0 & 0.667 & 0 & 0 & 0 & -0.083 & 0 & 0 & 0 \\ 0 & 0.667 & 0 & 1.417 & 0 & 0 & 0 & -0.667 & 0 & 0.667 & 0 & 0 \\ -0.667 & 0 & 0.667 & 0 & 2.667 & 0 & 0.667 & 0 & 0 & 0 & 0.667 & 0 \\ 0 & -0.667 & 0 & 0 & 0 & 1.417 & 0 & 0 & 0 & 0 & 0 & -0.833 \\ -0.444 & 0 & 0 & 0 & 0.667 & 0 & 0.889 & 0 & 0 & 0 & 0.667 & 0.667 \\ 0 & -0.444 & 0 & -0.667 & 0 & 0 & 0 & 0.528 & 0 & -0.667 & 0 & 0 \\ 0 & 0 & -0.083 & 0 & 0 & 0 & 0 & 0 & 0.528 & -0.667 & 0 & 0 \\ 0 & 0.667 & 0 & 0.667 & 0 & 0 & 0 & -0.667 & -0.667 & 2.667 & 0 & 0 \\ -0.667 & 0 & 0 & 0 & 0.667 & 0 & 0.667 & 0 & 0 & 0 & 1.417 & 0 \\ 0 & 0 & 0 & 0 & 0 & -0.083 & 0.667 & 0 & 0 & 0 & 0 & 1.417 \end{bmatrix}$$

7. Calculation of end moments and reactions:

$$\{M_1\} = \begin{Bmatrix} -0.045 \\ 55.725 \\ -1.543 \\ -2.820 \\ 4.346 \\ 73.232 \\ 0.045 \\ 4.275 \\ 1.543 \\ 2.820 \\ 0.284 \\ 3.943 \end{Bmatrix}, \quad \{M_2\} = \begin{Bmatrix} -0.045 \\ -4.275 \\ -1.543 \\ -2.820 \\ -0.284 \\ -3.943 \\ 0.045 \\ 4.275 \\ 1.543 \\ -10.004 \\ 0.420 \\ 3.943 \end{Bmatrix}, \quad \{M_3\} = \begin{Bmatrix} 0.045 \\ 4.275 \\ 1.543 \\ -5.374 \\ 0.420 \\ 3.807 \\ -0.045 \\ -4.275 \\ -1.543 \\ 10.004 \\ -0.420 \\ -3.943 \end{Bmatrix}$$

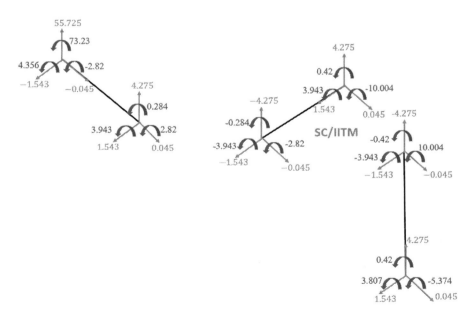

FIGURE 4.23 Member end reactions and moments.

Three-Dimensional Analysis of Space Frames

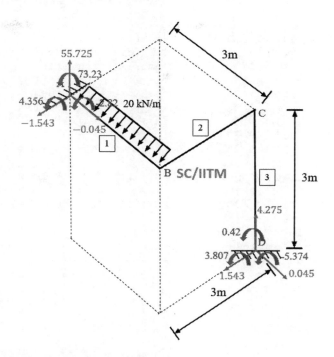

FIGURE 4.24 Final end reactions and moments.

The member end reactions and moments are shown in Figure 4.23 and the final end moments and reactions are shown in Figure 4.24.

MATLAB program:

```
%% 3D analysis of space frame
clc;
clear;
%% Input
n = 3; % number of members
EI = [1 1 1]; %Flexural rigidity
EIy = EI;
EIz = EI;
GI = [0.25 0.25 0.25].*EI; %Torsional constant
EA = [0.25 0.25 0.25].*EI; %Axial rigidity
L = [3 3 3]; % length in m
nj = n+1; % Number of Joints
codm = [0 0 0; 3 0 0; 3 0 -3; 3 -3 -3]; %Coordinate wrt X,Y.Z:
size=nj,3
dc = [1 0 0; 0 0 -1; 0 1 0]; % Direction cosines for each
member
tytr = [1 1 2]; % Type of transformation fo each member
psi = [0 0 90]; % Psi angle in degrees for each member
```

```
% C matrix
c1 = [1 0 0; 0 1 0; 0 0 1]; % C matrix for member 1
c2 = [0 0 -1; 0 1 0; 1 0 0]; % C matrix for member 2
c3 = [0 1 0; 0 0 1; 1 0 0]; % C matrix for member 3

uu = 12; % Number of unrestrained Degrees-of-freedom
ur = 12; % Number of restrained Degrees-of-freedom
uul = [1 2 3 4 5 6 7 8 9 10 11 12]; % global labels of
unrestrained dof
url = [13 14 15 16 17 18 19 20 21 22 23 24]; % global labels
of restrained dof
l1 = [13 14 15 16 17 18 1 2 3 4 5 6]; % Global labels for member 1
l2 = [1 2 3 4 5 6 7 8 9 10 11 12]; % Global labels for member 2
l3 = [19 20 21 22 23 24 7 8 9 10 11 12]; % Global labels for
member 3
l= [l1; l2; l3];
dof = uu + ur; % Degrees-of-freedom
Ktotal = zeros (dof);
fem1= [0; 30; 0; 0; 0; 15; 0; 30; 0; 0; 0; -15]; % Local Fixed
end moments of member 1
fem2= [0; 0; 0; 0; 0; 0; 0; 0; 0; 0; 0; 0]; % Local Fixed end
moments of member 2
fem3= [0; 0; 0; 0; 0; 0; 0; 0; 0; 0; 0; 0]; % Local Fixed end
moments of member 3

%% Transformation matrix
T1 = zeros(12);
T2 = zeros(12);
T3 = zeros(12);
for i = 1:3
    for j = 1:3
        T1(i,j)=c1(i,j);
        T1(i+3,j+3)=c1(i,j);
        T1(i+6,j+6)=c1(i,j);
        T1(i+9,j+9)=c1(i,j);
        T2(i,j)=c2(i,j);
        T2(i+3,j+3)=c2(i,j);
        T2(i+6,j+6)=c2(i,j);
        T2(i+9,j+9)=c2(i,j);
        T3(i,j)=c3(i,j);
        T3(i+3,j+3)=c3(i,j);
        T3(i+6,j+6)=c3(i,j);
        T3(i+9,j+9)=c3(i,j);
    end
end

%% Getting Type of transformation and Psi angle
for i = 1:n
    if tytr(i) ==1
        fprintf ('Member Number =');
        disp (i);
```

Three-Dimensional Analysis of Space Frames

```
            fprintf ('Type of transformation is Y-Z-X \n');
        else
            fprintf ('Member Number =');
            disp (i);
            fprintf ('Type of transformation is Z-Y-X \n');
        end
        fprintf ('Psi angle=');
        disp (psi(i));
end

%% Stiffness coefficients for each member
sc1 = EA./L;
sc2 = 6*EIz./(L.^2);
sc3 = 6*EIy./(L.^2);
sc4 = GI./L;
sc5 = 2*EIy./L;
sc6 = 12*EIz./(L.^3);
sc7 = 12*EIy./(L.^3);
sc8 = 2*EIz./L;

%% stiffness matrix 6 by 6
for i = 1:n
        Knew = zeros (dof);
        k1 = [sc1(i); 0; 0; 0; 0; 0; -sc1(i); 0; 0; 0; 0; 0];
        k2 = [0; sc6(i); 0; 0; 0; sc2(i); 0; -sc6(i); 0; 0; 0; sc2(i)];
        k3 = [0; 0; sc7(i); 0; -sc3(i); 0; 0; 0; -sc7(i); 0; -sc3(i); 0];
        k4 = [0; 0; 0; sc4(i); 0; 0; 0; 0; 0; -sc4(i); 0; 0];
        k5 = [0; 0; -sc3(i); 0; (2*sc5(i)); 0; 0; 0; sc3(i); 0; sc5(i); 0];
        k6 = [0; sc2(i); 0; 0; 0; (2*sc8(i)); 0; -sc2(i); 0; 0; 0; sc8(i)];
        k7 = -k1;
        k8 = -k2;
        k9 = -k3;
        k10 = -k4;
        k11 = [0; 0; -sc3(i); 0; sc5(i); 0; 0; 0; sc3(i); 0; (2*sc5(i)); 0];
        k12 = [0; sc2(i); 0; 0; 0; sc8(i); 0; -sc2(i); 0; 0; 0; (2*sc8(i))];
        K = [k1 k2 k3 k4 k5 k6 k7 k8 k9 k10 k11 k12];
        fprintf ('Member Number =');
        disp (i);
        fprintf ('Local Stiffness matrix of member, [K] = \n');
        disp (K);
        if i == 1
            T = T1;
        elseif i == 2
            T = T2;
        else
```

```
            T = T3;
        end
        Ttr = T';
        Kg = Ttr*K*T;
        fprintf ('Transformation matrix, [T] = \n');
        disp (T);
        fprintf ('Global Matrix, [K global] = \n');
        disp (Kg);
            for p = 1:12
            for q = 1:12
                Knew((l(i,p)),(l(i,q))) =Kg(p,q);
            end
        end
        Ktotal = Ktotal + Knew;
        if i == 1
            Tt1= T;
            Kg1=Kg;
            fembar1= Tt1'*fem1;
        elseif i == 2
            Tt2 = T;
            Kg2 = Kg;
            fembar2= Tt2'*fem2;
        else
            Tt3 = T;
            Kg3 = Kg;
            fembar3= Tt3'*fem3;
        end
end
fprintf ('Stiffness Matrix of complete structure, [Ktotal] = \n');
disp (Ktotal);
Kunr = zeros(12);
for x=1:uu
    for y=1:uu
        Kunr(x,y)= Ktotal(x,y);
    end
end
fprintf ('Unrestrained Stiffness sub-matrix, [Kuu] = \n');
disp (Kunr);
KuuInv= inv(Kunr);
fprintf ('Inverse of Unrestrained Stiffness sub-matrix,
[KuuInverse] = \n');
disp (KuuInv);

%% Creation of joint load vector
jl= [0; -30; 0; 0; 0; 15; 0; 0; 0; 0; 0; 0; 0; -30; 0; 0; 0;
-15; 0; 0; 0; 0; 0; 0]; % values given in kN or kNm
jlu = jl(1:12,1); % load vector in unrestrained dof
delu = KuuInv*jlu;
fprintf ('Joint Load vector, [Jl] = \n');
disp (jl);
fprintf ('Unrestrained displacements, [DelU] = \n');
disp (delu);
```

Three-Dimensional Analysis of Space Frames

```
delr = [0; 0; 0; 0; 0; 0; 0; 0; 0; 0; 0; 0];
del = zeros (dof,1);
del = [delu; delr];
deli= zeros (12,1);
for i = 1:n
    for p = 1:12
        deli(p,1) = del((l(i,p)),1) ;
    end
    if i == 1
            delbar1 = deli;
            mbar1= (Kg1 * delbar1)+fembar1;
            fprintf ('Member Number =');
            disp (i);
            fprintf ('Global displacement matrix [DeltaBar] = \n');
            disp (delbar1);
            fprintf ('Global End moment matrix [MBar] = \n');
            disp (mbar1);
      elseif i == 2
            delbar2 = deli;
            mbar2= (Kg2 * delbar2)+fembar2;
            fprintf ('Member Number =');
            disp (i);
            fprintf ('Global displacement matrix [DeltaBar] = \n');
            disp (delbar2);
            fprintf ('Global End moment matrix [MBar] = \n');
            disp (mbar2);
      else
            delbar3 = deli;
            mbar3= (Kg3 * delbar3)+fembar3;
            fprintf ('Member Number =');
            disp (i);
            fprintf ('Global displacement matrix [DeltaBar] = \n');
            disp (delbar3);
            fprintf ('Global End moment matrix [MBar] = \n');
            disp (mbar3);
      end
end

%% check
mbar = [mbar1'; mbar2'; mbar3'];
jf = zeros(dof,1);
for a=1:n
    for b=1:12 % size of k matrix
        d = l(a,b);
        jfnew = zeros(dof,1);
        jfnew(d,1)=mbar(a,b);
        jf=jf+jfnew;
    end
end
fprintf ('Joint forces = \n');
disp (jf);
```

EXAMPLE 4.4:

Analyze the three-dimensional space frame, as shown in Figure 4.25, using the stiffness method of analysis.

SOLUTION:

Mark the local and reference axes system, as shown in Figure 4.26.

1. *Coordinates of joints:*

Joint	X	Y	Z
A	0	6	0
B	4	6	0
C	7	6	0
D	0	0	0
E	4	0	0
F	7	0	0

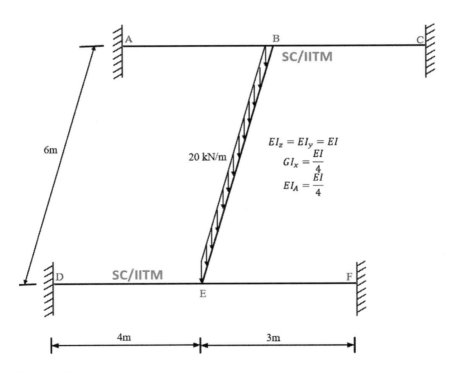

FIGURE 4.25 Space frame example.

Three-Dimensional Analysis of Space Frames

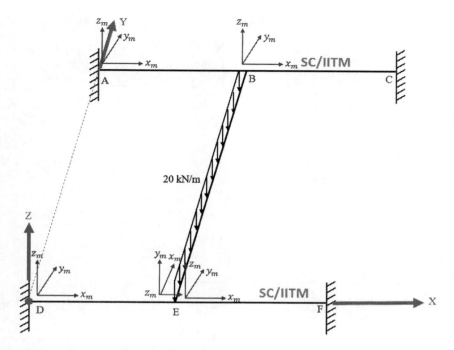

FIGURE 4.26 Local and reference axes system.

2. ψ angle:

Member	Length (m)	Joint j	Joint k	C_x	C_y	C_z	Type of Transformation	ψ Angle (degrees)
AB	4	A	B	1	0	0	Y-Z-X	$\psi_y = 0$
BC	3	B	C	1	0	0	Y-Z-X	$\psi_y = 0$
DE	4	D	E	1	0	0	Y-Z-X	$\psi_y = 0$
EF	3	E	F	1	0	0	Y-Z-X	$\psi_y = 0$
EB	6	E	B	0	1	0	Z-Y-X	$\psi_z = 90$

3. *Marking unrestrained and restrained degrees-of-freedom:*

The unrestrained and restrained degrees-of-freedom are marked, as shown in Figure 4.27.

Unrestrained degrees-of-freedom: 12 ($\bar{\delta}_1, \bar{\delta}_2, \bar{\delta}_3, \bar{\theta}_4, \bar{\theta}_5, \bar{\theta}_6, \bar{\delta}_7, \bar{\delta}_8, \bar{\delta}_9, \bar{\theta}_{10}, \bar{\theta}_{11}, \bar{\theta}_{12}$)

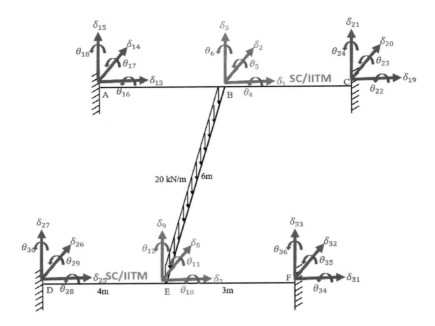

FIGURE 4.27 Unrestrained and restrained degrees-of-freedom.

Restrained degrees-of-freedom:

$$24 \begin{matrix} (\bar{\delta}_{13}, \bar{\delta}_{14}, \bar{\delta}_{15}, \bar{\theta}_{16}, \bar{\theta}_{17}, \bar{\theta}_{18}, \bar{\delta}_{19}, \bar{\delta}_{20}, \bar{\delta}_{21}, \bar{\theta}_{22}, \bar{\theta}_{23}, \bar{\theta}_{24}) \\ (\bar{\delta}_{25}, \bar{\delta}_{26}, \bar{\delta}_{27}, \bar{\theta}_{28}, \bar{\theta}_{29}, \bar{\theta}_{30}, \bar{\delta}_{31}, \bar{\delta}_{32}, \bar{\delta}_{33}, \bar{\theta}_{34}, \bar{\theta}_{35}, \bar{\theta}_{36}) \end{matrix}$$

4. *Estimation of transformation matrix:*

We already know that,

$$C = \begin{bmatrix} C_{11} & C_{12} & C_{13} \\ C_{21} & C_{22} & C_{23} \\ C_{31} & C_{32} & C_{33} \end{bmatrix}$$

The direction cosine matrices for all the three members are,

$$C_{AB} = \begin{bmatrix} 1 & 0 & 0 \\ 0 & 1 & 0 \\ 0 & 0 & 1 \end{bmatrix}, C_{EB} = \begin{bmatrix} 0 & 1 & 0 \\ 0 & 0 & 1 \\ 1 & 0 & 0 \end{bmatrix}$$

$$C_{AB} = C_{BC} = C_{DE} = C_{EF}$$

Three-Dimensional Analysis of Space Frames

Now, the transformation matrix is given by,

$$[T_i] = \begin{bmatrix} [C_i] & [0] & [0] & [0] \\ [0] & [C_i] & [0] & [0] \\ [0] & [0] & [C_i] & [0] \\ [0] & [0] & [0] & [C_i] \end{bmatrix}_{12 \times 12}$$

Global labels:

AB = [13, 14, 15, 16, 17, 18, 1, 2, 3, 4, 5, 6]
BC = [1, 2, 3, 4, 5, 6, 19, 20, 21, 22, 23, 24]
DE = [25, 26, 27, 28, 29, 30, 7, 8, 9, 10, 11, 12]
EF = [7, 8, 9, 10, 11, 12, 31, 32, 33, 34, 35, 36]
EB = [7, 8, 9, 10, 11, 12, 1, 2, 3, 4, 5, 6]

5. *Fixed end moments and joint load vector:*

Load is acting only on the member EB. Hence, the fixed end moments along the other two members remains zero.
Member EB:

$$V_A = \frac{20 \times 6}{2} = 60\,\text{kN}$$

$$V_B = 60\,\text{kN}$$

$$M_{EB} = \frac{20 \times 6^2}{12} = 60\,\text{kNm}$$

$$M_{BE} = 60\,\text{kNm}$$

$$(\text{FEM})_{EB} = \begin{Bmatrix} 0 \\ 0 \\ 60 \\ -60 \\ 0 \\ 0 \\ 0 \\ 0 \\ 60 \\ 60 \\ 0 \\ 0 \end{Bmatrix}$$

The joint load vector is the reversal of the fixed end moments. The transpose of the joint load vector is given as follows:

$$[J_L]^T = \{0 \quad 0 \quad -60 \quad 60 \quad 0 \quad 0 \quad 0 \quad 0 \quad -60 \quad -60 \quad [0]\}_{36\times 1}$$

6. *Stiffness matrix:*

$$[K_{AB}] = EI \begin{bmatrix}
& \text{⑬} & \text{⑭} & \text{⑮} & \text{⑯} & \text{⑰} & \text{⑱} & \text{①} & \text{②} & \text{③} & \text{④} & \text{⑤} & \text{⑥} \\
0.063 & 0 & 0 & 0 & 0 & 0 & -0.063 & 0 & 0 & 0 & 0 & 0 \\
0 & 0.188 & 0 & 0 & 0 & 0.375 & 0 & -0.188 & 0 & 0 & 0 & 0.375 \\
0 & 0 & 0.188 & 0 & -0.375 & 0 & 0 & 0 & -0.188 & 0 & -0.375 & 0 \\
0 & 0 & 0 & 0.063 & 0 & 0 & 0 & 0 & 0 & -0.063 & 0 & 0 \\
0 & 0 & -0.375 & 0 & 1 & 0 & 0 & 0 & 0.375 & 0 & 0.5 & 0 \\
0 & 0.375 & 0 & 0 & 0 & 1 & 0 & -0.375 & 0 & 0 & 0 & 0.5 \\
-0.063 & 0 & 0 & 0 & 0 & 0 & 0.063 & 0 & 0 & 0 & 0 & 0 \\
0 & 0-0.188 & 0 & 0 & 0 & -0.375 & 0 & 0.88 & 0 & 0 & 0 & -0.375 \\
0 & 0 & -0.188 & 0 & 0.375 & 0 & 0 & 0 & 0.188 & 0 & 0.75 & 0 \\
0 & 0 & 0 & -0.063 & 0 & 0 & 0 & 0 & 0 & 0.063 & 0 & 0 \\
0 \cdot & 0 & -0.375 & 0 & 0.5 & 0 & 0 & 0 & 0.375 & 0 & 1 & 0 \\
0 & 0.375 & 0 & 0 & 0 & 0.5 & 0 & -0.375 & 0 & 0 & 0 & 1
\end{bmatrix} \begin{matrix} \text{⑬} \\ \text{⑭} \\ \text{⑮} \\ \text{⑯} \\ \text{⑰} \\ \text{⑱} \\ \text{①} \\ \text{②} \\ \text{③} \\ \text{④} \\ \text{⑤} \\ \text{⑥} \end{matrix}$$

$$[\bar{K}_{AB}] = [K_{AB}]$$

$$[K_{BC}] = EI \begin{bmatrix}
& \text{①} & \text{②} & \text{③} & \text{④} & \text{⑤} & \text{⑥} & \text{⑲} & \text{⑳} & \text{㉑} & \text{㉒} & \text{㉓} & \text{㉔} \\
0.083 & 0 & 0 & 0 & 0 & 0 & -0.083 & 0 & 0 & 0 & 0 & 0 \\
0 & 0.444 & 0 & 0 & 0 & 0.667 & 0 & -0.444 & 0 & 0 & 0 & 0.667 \\
0 & 0 & 0.444 & 0 & 0-0.667 & 0 & 0 & 0 & -0.444 & 0 & -0.667 & 0 \\
0 & 0 & 0 & 0.083 & 0 & 0 & 0 & 0 & 0 & -0.083 & 0 & 0 \\
0 & 0 & -0.667 & 0 & 1.333 & 0 & 0 & 0 & 0.667 & 0 & 0.667 & 0 \\
0 & 0.667 & 0 & 0 & 0 & 1.333 & 0 & -0.667 & 0 & 0 & 0 & 0.667 \\
-0.083 & 0 & 0 & 0 & 0 & 0 & 0.083 & 0 & 0 & 0 & 0 & 0 \\
0 & -0.444 & 0 & 0 & 0 & -0.667 & 0 & 0.444 & 0 & 0 & 0 & -0.667 \\
0 & 0 & -0.444 & 0 & 0.667 & 0 & 0 & 0 & 0.444 & 0 & 0.667 & 0 \\
0 & 0 & 0 & -0.083 & 0 & 0 & 0 & 0 & 0 & 0.083 & 0 & 0 \\
0 & 0 & -0.667 & 0 & 0.667 & 0 & 0 & 0 & 0.667 & 0 & 1.333 & 0 \\
0 & 0.667 & 0 & 0 & 0 & 0.667 & 0 & -0.667 & 0 & 0 & 0 & 1.333
\end{bmatrix} \begin{matrix} \text{①} \\ \text{②} \\ \text{③} \\ \text{④} \\ \text{⑤} \\ \text{⑥} \\ \text{⑲} \\ \text{⑳} \\ \text{㉑} \\ \text{㉒} \\ \text{㉓} \\ \text{㉔} \end{matrix}$$

$$[\bar{K}_{BC}] = [K_{BC}]$$

$$[K_{DE}] = EI \begin{bmatrix}
& \text{㉕} & \text{㉖} & \text{㉗} & \text{㉘} & \text{㉙} & \text{㉚} & \text{⑦} & \text{⑧} & \text{⑨} & \text{⑩} & \text{⑪} & \text{⑫} \\
0.083 & 0 & 0 & 0 & 0 & 0 & -0.083 & 0 & 0 & 0 & 0 & 0 \\
0 & 0.444 & 0 & 0 & 0 & 0.667 & 0 & -0.444 & 0 & 0 & 0 & 0.667 \\
0 & 0 & 0.444 & 0 & 0-0.667 & 0 & 0 & 0 & -0.444 & 0 & -0.667 & 0 \\
0 & 0 & 0 & 0.083 & 0 & 0 & 0 & 0 & 0 & -0.083 & 0 & 0 \\
0 & 0 & -0.667 & 0 & 1.333 & 0 & 0 & 0 & 0.667 & 0 & 0.667 & 0 \\
0 & 0.667 & 0 & 0 & 0 & 1.333 & 0 & -0.667 & 0 & 0 & 0 & 0.667 \\
-0.083 & 0 & 0 & 0 & 0 & 0 & 0.083 & 0 & 0 & 0 & 0 & 0 \\
0 & -0.444 & 0 & 0 & 0 & -0.667 & 0 & 0.444 & 0 & 0 & 0 & -0.667 \\
0 & 0 & -0.444 & 0 & 0.667 & 0 & 0 & 0 & 0.444 & 0 & 0.667 & 0 \\
0 & 0 & 0 & -0.083 & 0 & 0 & 0 & 0 & 0 & 0.083 & 0 & 0 \\
0 & 0 & -0.667 & 0 & 0.667 & 0 & 0 & 0 & 0.667 & 0 & 1.333 & 0 \\
0 & 0.667 & 0 & 0 & 0 & 0.667 & 0 & -0.667 & 0 & 0 & 0 & 1.333
\end{bmatrix} \begin{matrix} \text{㉕} \\ \text{㉖} \\ \text{㉗} \\ \text{㉘} \\ \text{㉙} \\ \text{㉚} \\ \text{⑦} \\ \text{⑧} \\ \text{⑨} \\ \text{⑩} \\ \text{⑪} \\ \text{⑫} \end{matrix}$$

Three-Dimensional Analysis of Space Frames

$$[\bar{K}_{DE}] = [K_{DE}]$$

$$[K_{EF}] = EI\begin{bmatrix}
0.083 & 0 & 0 & 0 & 0 & 0 & -0.083 & 0 & 0 & 0 & 0 & 0 \\
0 & 0.444 & 0 & 0 & 0 & 0.667 & 0 & -0.444 & 0 & 0 & 0 & 0.667 \\
0 & 0 & 0.444 & 0 & 0 & -0.667 & 0 & 0 & 0 & -0.444 & 0 & -0.667 & 0 \\
0 & 0 & 0 & 0.083 & 0 & 0 & 0 & 0 & 0 & -0.083 & 0 & 0 \\
0 & 0 & -0.667 & 0 & 1.333 & 0 & 0 & 0 & 0.667 & 0 & 0.667 & 0 \\
0 & 0.667 & 0 & 0 & 0 & 1.333 & 0 & -0.667 & 0 & 0 & 0 & 0.667 \\
-0.083 & 0 & 0 & 0 & 0 & 0 & 0.083 & 0 & 0 & 0 & 0 & 0 \\
0 & -0.444 & 0 & 0 & 0 & -0.667 & 0 & 0.444 & 0 & 0 & 0 & -0.667 \\
0 & 0 & -0.444 & 0 & 0.667 & 0 & 0 & 0 & 0.444 & 0 & 0.667 & 0 \\
0 & 0 & 0 & -0.083 & 0 & 0 & 0 & 0 & 0 & 0.083 & 0 & 0 \\
0 & 0 & -0.667 & 0 & 0.667 & 0 & 0 & 0 & 0.667 & 0 & 1.333 & 0 \\
0 & 0.667 & 0 & 0 & 0 & 0.667 & 0 & -0.667 & 0 & 0 & 0 & 1.333
\end{bmatrix}\begin{matrix} 7 \\ 8 \\ 9 \\ 10 \\ 11 \\ 12 \\ 31 \\ 32 \\ 33 \\ 34 \\ 35 \\ 36 \end{matrix}$$

columns: 7, 8, 9, 10, 11, 12, 31, 32, 33, 34, 35, 36

$$[\bar{K}_{EF}] = [K_{EF}]$$

$$[K_{EF}] = EI\begin{bmatrix}
0.083 & 0 & 0 & 0 & 0 & 0 & -0.083 & 0 & 0 & 0 & 0 & 0 \\
0 & 0.444 & 0 & 0 & 0 & 0.667 & 0 & -0.444 & 0 & 0 & 0 & 0.667 \\
0 & 0 & 0.444 & 0 & 0 & -0.667 & 0 & 0 & 0 & -0.444 & 0 & -0.667 & 0 \\
0 & 0 & 0 & 0.083 & 0 & 0 & 0 & 0 & 0 & -0.083 & 0 & 0 \\
0 & 0 & -0.667 & 0 & 1.333 & 0 & 0 & 0 & 0.667 & 0 & 0.667 & 0 \\
0 & 0.667 & 0 & 0 & 0 & 1.333 & 0 & -0.667 & 0 & 0 & 0 & 0.667 \\
-0.083 & 0 & 0 & 0 & 0 & 0 & 0.083 & 0 & 0 & 0 & 0 & 0 \\
0 & -0.444 & 0 & 0 & 0 & -0.667 & 0 & 0.444 & 0 & 0 & 0 & -0.667 \\
0 & 0 & -0.444 & 0 & 0.667 & 0 & 0 & 0 & 0.444 & 0 & 0.667 & 0 \\
0 & 0 & 0 & -0.083 & 0 & 0 & 0 & 0 & 0 & 0.083 & 0 & 0 \\
0 & 0 & -0.667 & 0 & 0.667 & 0 & 0 & 0 & 0.667 & 0 & 1.333 & 0 \\
0 & 0.667 & 0 & 0 & 0 & 0.667 & 0 & -0.667 & 0 & 0 & 0 & 1.333
\end{bmatrix}\begin{matrix} 7 \\ 8 \\ 9 \\ 10 \\ 11 \\ 12 \\ 1 \\ 2 \\ 3 \\ 4 \\ 5 \\ 6 \end{matrix}$$

columns: 7, 8, 9, 10, 11, 12, 1, 2, 3, 4, 5, 6

$$[\bar{K}_{EB}] = [K_{EB}]$$

The total stiffness matrix can be formed by assembling the global stiffness matrices of all the members, from which the unrestrained stiffness matrix can be partitioned.

$$[K_{uu}] = EI\begin{bmatrix}
0.188 & 0 & 0 & 0 & 0 & 0 & -0.042 & 0 & 0 & 0 & 0 & 0 \\
0 & 0.688 & 0 & 0 & 0 & 0.125 & 0 & -0.056 & 0 & 0 & 0 & -0.167 \\
0 & 0 & 0.688 & 0 & -0.125 & 0 & 0 & 0 & -0.056 & 0 & 0.167 & 0 \\
0 & 0 & 0 & 0.188 & 0 & 0 & 0 & 0 & 0 & -0.042 & 0 & 0 \\
0 & 0 & -0.125 & 0 & 3 & 0 & 0 & 0 & -0.167 & 0 & 0.333 & 0 \\
0 & 0.125 & 0 & 0 & 0 & 3 & 0 & 0.167 & 0 & 0 & 0 & 0.333 \\
-0.042 & 0 & 0 & 0 & 0 & 0 & 0.188 & 0 & 0 & 0 & 0 & 0 \\
0 & -0.056 & 0 & 0 & 0 & 0.167 & 0 & 0.688 & 0 & 0 & 0 & 0.458 \\
0 & 0 & -0.056 & 0 & -0.167 & 0 & 0 & 0 & 0.688 & 0 & -0.458 & 0 \\
0 & 0 & 0 & -0.042 & 0 & 0 & 0 & 0 & 0 & 0.188 & 0 & 0 \\
0 & 0 & 0.167 & 0 & 0.333 & 0 & 0 & 0 & -0.458 & 0 & 3 & 0 \\
0 & -0.167 & 0 & 0 & 0 & 0.333 & 0 & 0.458 & 0 & 0 & 0 & 3
\end{bmatrix}\begin{matrix} 1 \\ 2 \\ 3 \\ 4 \\ 5 \\ 6 \\ 7 \\ 8 \\ 9 \\ 10 \\ 11 \\ 12 \end{matrix}$$

columns: 1, 2, 3, 4, 5, 6, 7, 8, 9, 10, 11, 12

198   Advanced Structural Analysis with MATLAB®

7. *Calculation of end moments and reactions:*

$$\{M_1\} = \begin{Bmatrix} 0 \\ 0 \\ 21.028 \\ -16.364 \\ -39.898 \\ 0 \\ 0 \\ 0 \\ -21.028 \\ 16.364 \\ -44.215 \\ 0 \end{Bmatrix}, \{M_2\} = \begin{Bmatrix} 0 \\ 0 \\ -36.412 \\ 21.818 \\ 51.739 \\ 0 \\ 0 \\ 0 \\ 36.412 \\ -21.818 \\ 57.496 \\ 0 \end{Bmatrix}, \{M_3\} = \begin{Bmatrix} 0 \\ 0 \\ 22.978 \\ 16.364 \\ -43.564 \\ 0 \\ 0 \\ 0 \\ -22.978 \\ -16.363 \\ -48.348 \\ 0 \end{Bmatrix}, \{M_4\} = \begin{Bmatrix} 0 \\ 0 \\ -39.582 \\ -21.818 \\ 56.183 \\ 0 \\ 0 \\ 0 \\ 39.582 \\ 21.818 \\ 62.562 \\ 0 \end{Bmatrix}, \{M_5\} = \begin{Bmatrix} 0 \\ 0 \\ 62.560 \\ 38.182 \\ -7.835 \\ 0 \\ 0 \\ 0 \\ 57.440 \\ -38.182 \\ -7.524 \\ 0 \end{Bmatrix}$$

The member end forces and moments are shown in Figure 4.28. The final end moments and reactions are shown in Figure 4.29.

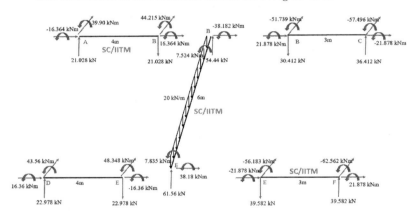

FIGURE 4.28 Member end forces and moments.

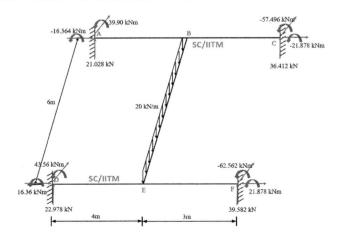

FIGURE 4.29 Member end forces and moments.

Three-Dimensional Analysis of Space Frames

MATLAB® program:

```
%% 3D analysis of space frame
clc;
clear;
%% Input
n = 5; % number of members
EI = [1 1 1 1 1]; %Flexural rigidity
EIy = EI;
EIz = EI;
GI = [0.25 0.25 0.25 0.25 0.25].*EI; %Torsional constant
EA = [0.25 0.25 0.25 0.25 0.25].*EI; %Axial rigidity
L = [4 3 4 3 6]; % length in m
nj = n+1; % Number of Joints
codm = [0 6 0; 4 6 0; 7 6 0; 0 0 0; 4 0 0; 7 0 0]; %Coordinate
wrt X,Y.Z: size=nj,3
dc = [1 0 0; 1 0 0; 1 0 0; 1 0 0; 0 1 0]; % Direction cosines
for each member
tytr = [1 1 1 1 2]; % Type of transformation fo each member
psi = [0 0 0 0 90]; % Psi angle in degrees for each member

% C matrix
c1 = [1 0 0; 0 1 0; 0 0 1]; % C matrix for member 1
c2 = [1 0 0; 0 1 0; 0 0 1]; % C matrix for member 2
c3 = [1 0 0; 0 1 0; 0 0 1]; % C matrix for member 3
c4 = [1 0 0; 0 1 0; 0 0 1]; % C matrix for member 4
c5 = [0 1 0; 0 0 1; 1 0 0]; % C matrix for member 5

uu = 12; % Number of unrestrained Degrees-of-freedom
ur = 24; % Number of restrained Degrees-of-freedom
uul = [1 2 3 4 5 6 7 8 9 10 11 12]; % global labels of
unrestrained dof
url = [13 14 15 16 17 18 19 20 21 22 23 24 25 26 27 28 29 30
31 32 33 34 35 36]; % global labels of restrained dof

l1 = [13 14 15 16 17 18 1 2 3 4 5 6]; % Global labels for
member 1
l2 = [1 2 3 4 5 6 19 20 21 22 23 24]; % Global labels for
member 2
l3 = [25 26 27 28 29 30 7 8 9 10 11 12]; % Global labels for
member 3
l4 = [7 8 9 10 11 12 31 32 33 34 35 36]; % Global labels for
member 4
l5 = [7 8 9 10 11 12 1 2 3 4 5 6]; % Global labels for member
5
l= [l1; l2; l3; l4; l5];
dof = uu + ur; % Degrees-of-freedom
Ktotal = zeros (dof);

fem1= [0; 0; 0; 0; 0; 0; 0; 0; 0; 0; 0; 0]; % Local Fixed end
moments of member 1
fem2= [0; 0; 0; 0; 0; 0; 0; 0; 0; 0; 0; 0]; % Local Fixed end
moments of member 2
```

```
fem3= [0; 0; 0; 0; 0; 0; 0; 0; 0; 0; 0; 0]; % Local Fixed end
moments of member 3
fem4= [0; 0; 0; 0; 0; 0; 0; 0; 0; 0; 0; 0]; % Local Fixed end
moments of member 4
fem5= [0; 0; 60; 60; 0; 0; 0; 0; 60; -60; 0; 0]; % Local Fixed
end moments of member 5

%% Transformation matrix
T1 = zeros(12);
T2 = zeros(12);
T3 = zeros(12);
T4 = zeros(12);
T5 = zeros(12);
for i = 1:3
    for j = 1:3
        T1(i,j)=c1(i,j);
        T1(i+3,j+3)=c1(i,j);
        T1(i+6,j+6)=c1(i,j);
        T1(i+9,j+9)=c1(i,j);
        T2(i,j)=c2(i,j);
        T2(i+3,j+3)=c2(i,j);
        T2(i+6,j+6)=c2(i,j);
        T2(i+9,j+9)=c2(i,j);
        T3(i,j)=c3(i,j);
        T3(i+3,j+3)=c3(i,j);
        T3(i+6,j+6)=c3(i,j);
        T3(i+9,j+9)=c3(i,j);
        T4(i,j)=c4(i,j);
        T4(i+3,j+3)=c4(i,j);
        T4(i+6,j+6)=c4(i,j);
        T4(i+9,j+9)=c4(i,j);
        T5(i,j)=c5(i,j);
        T5(i+3,j+3)=c5(i,j);
        T5(i+6,j+6)=c5(i,j);
        T5(i+9,j+9)=c5(i,j);
    end
end

%% Getting Type of transformation and Psi angle
for i = 1:n
    if tytr(i) ==1
        fprintf ('Member Number =');
        disp (i);
        fprintf ('Type of transformation is Y-Z-X \n');
    else
        fprintf ('Member Number =');
        disp (i);
        fprintf ('Type of transformation is Z-Y-X \n');
    end
    fprintf ('Psi angle=');
    disp (psi(i));
end
```

Three-Dimensional Analysis of Space Frames

```
%% Stiffness coefficients for each member
sc1 = EA./L;
sc2 = 6*EIz./(L.^2);
sc3 = 6*EIy./(L.^2);
sc4 = GI./L;
sc5 = 2*EIy./L;
sc6 = 12*EIz./(L.^3);
sc7 = 12*EIy./(L.^3);
sc8 = 2*EIz./L;

%% stiffness matrix 6 by 6
for i = 1:n
        Knew = zeros (dof);
        k1 = [sc1(i); 0; 0; 0; 0; 0; -sc1(i); 0; 0; 0; 0; 0];
        k2 = [0; sc6(i); 0; 0; 0; sc2(i); 0; -sc6(i); 0; 0; 0; sc2(i)];
        k3 = [0; 0; sc7(i); 0; -sc3(i); 0; 0; 0; -sc7(i); 0; -sc3(i); 0];
        k4 = [0; 0; 0; sc4(i); 0; 0; 0; 0; 0; -sc4(i); 0; 0];
        k5 = [0; 0; -sc3(i); 0; (2*sc5(i)); 0; 0; 0; sc3(i); 0; sc5(i); 0];
        k6 = [0; sc2(i); 0; 0; 0; (2*sc8(i)); 0; -sc2(i); 0; 0; 0; sc8(i)];
        k7 = -k1;
        k8 = -k2;
        k9 = -k3;
        k10 = -k4;
        k11 = [0; 0; -sc3(i); 0; sc5(i); 0; 0; 0; sc3(i); 0; (2*sc5(i)); 0];
        k12 = [0; sc2(i); 0; 0; 0; sc8(i); 0; -sc2(i); 0; 0; 0; (2*sc8(i))];
        K = [k1 k2 k3 k4 k5 k6 k7 k8 k9 k10 k11 k12];
        fprintf ('Member Number =');
        disp (i);
        fprintf ('Local Stiffness matrix of member, [K] = \n');
        disp (K);
        if i == 1
            T = T1;
        elseif i == 2
            T = T2;
        else
            T = T3;
        end
        Ttr = T';
        Kg = Ttr*K*T;
        fprintf ('Global Matrix, [K global] = \n');
        disp (Kg);

        for p = 1:12
            for q = 1:12
                Knew((l(i,p)),(l(i,q))) =Kg(p,q);
            end
        end
```

```
            Ktotal = Ktotal + Knew;
            if i == 1
                Tt1= T;
                Kg1=Kg;
                fembar1= Tt1'*fem1;
            elseif i == 2
                Tt2 = T;
                Kg2 = Kg;
                fembar2= Tt2'*fem2;
            elseif i ==3
                Tt3 = T;
                Kg3 = Kg;
                fembar3= Tt3'*fem3;
            elseif i ==4
                Tt4 = T;
                Kg4 = Kg;
                fembar4= Tt4'*fem4;
             else
                Tt5 = T;
                Kg5 = Kg;
                fembar5= Tt5'*fem5;
            end
end
fprintf ('Stiffness Matrix of complete structure, [Ktotal] = \n');
disp (Ktotal);
Kunr = zeros(12);
for x=1:uu
    for y=1:uu
        Kunr(x,y)= Ktotal(x,y);
    end
end
fprintf ('Unrestrained Stiffness sub-matrix, [Kuu] = \n');
disp (Kunr);
KuuInv= inv(Kunr);
fprintf ('Inverse of Unrestrained Stiffness sub-matrix,
[KuuInverse] = \n');
disp (KuuInv);

%% Creation of joint load vector
jl= [0; 0; -60; 60; 0; 0; 0; 0; -60; -60; 0; 0; 0; 0; 0; 0; 0;
0; 0; 0; 0; 0; 0; 0; 0; 0; 0; 0; 0; 0; 0; 0; 0; 0]; %
values given in kN or kNm
jlu = jl(1:12,1); % load vector in unrestrained dof
delu = KuuInv*jlu;
fprintf ('Joint Load vector, [Jl] = \n');
disp (jl');
fprintf ('Unrestrained displacements, [DelU] = \n');
disp (delu');
delr = [0; 0; 0; 0; 0; 0; 0; 0; 0; 0; 0; 0; 0; 0; 0; 0; 0; 0;
0; 0; 0; 0; 0; 0];
del = zeros (dof,1);
del = [delu; delr];
```

Three-Dimensional Analysis of Space Frames

```
deli= zeros (12,1);
for i = 1:n
   for p = 1:12
       deli(p,1) = del((l(i,p)),1) ;
   end
   if i == 1
         delbar1 = deli;
         mbar1= (Kg1 * delbar1)+fembar1;
         fprintf ('Member Number =');
         disp (i);
         fprintf ('Global displacement matrix [DeltaBar] = \n');
         disp (delbar1');
         fprintf ('Global End moment matrix [MBar] = \n');
         disp (mbar1');
      elseif i == 2
         delbar2 = deli;
         mbar2= (Kg2 * delbar2)+fembar2;
         fprintf ('Member Number =');
         disp (i);
         fprintf ('Global displacement matrix [DeltaBar] = \n');
         disp (delbar2');
         fprintf ('Global End moment matrix [MBar] = \n');
         disp (mbar2');
      elseif i ==3
         delbar3 = deli;
         mbar3= (Kg3 * delbar3)+fembar3;
         fprintf ('Member Number =');
         disp (i);
         fprintf ('Global displacement matrix [DeltaBar] = \n');
         disp (delbar3');
         fprintf ('Global End moment matrix [MBar] = \n');
         disp (mbar3');
      elseif i ==4
         delbar4 = deli;
         mbar4= (Kg4 * delbar4)+fembar4;
         fprintf ('Member Number =');
         disp (i);
         fprintf ('Global displacement matrix [DeltaBar] = \n');
         disp (delbar4');
         fprintf ('Global End moment matrix [MBar] = \n');
         disp (mbar4');
      else
         delbar5 = deli;
         mbar5= (Kg5 * delbar5)+fembar5;
         fprintf ('Member Number =');
         disp (i);
         fprintf ('Global displacement matrix [DeltaBar] = \n');
         disp (delbar5');
         fprintf ('Global End moment matrix [MBar] = \n');
         disp (mbar5');
    end
end
```

```
%% check
mbar = [mbar1'; mbar2'; mbar3'; mbar4'; mbar5'];
jf = zeros(dof,1);
for a=1:n
    for b=1:12 % size of k matrix
        d = l(a,b);
        jfnew = zeros(dof,1);
        jfnew(d,1)=mbar(a,b);
        jf=jf+jfnew;
    end
end
fprintf ('Joint forces = \n');
disp (jf');
```

5 Analysis of Special Members

5.1 THREE-DIMENSIONAL ANALYSIS OF TRUSS STRUCTURES

There are several general assumptions made in the two-dimensional analysis of truss structures. Joints are assumed to be pinned connections, which is one of the basic assumptions made in the analysis of planar frames. It is also valid for three-dimensional truss systems. A beam element developed earlier, will be used here with a small modification. Assume that the beam element has spherical hinges at both ends. The consequence of this assumption is that the beam can freely rotate about any axes. Thus, the end rotations will be zero. For the beam element in three-dimensional analysis, the number of degrees-of-freedom is 12. Each end will have three translations and three rotations. But, in this case, the beam element is restrained with spherical hinges. Thus, the beam element will have three displacement components at each end of the member. The truss member can resist only axial deformation and axial forces, which makes the stiffness matrix of order 6×6.

Consider a typical truss member arbitrarily oriented, as shown in Figure 5.1. The local axes system and the reference axes system for the member are marked. The degrees-of-freedom in the local and reference axes system are also marked at both ends of the truss member. The corresponding axial forces are mentioned in brackets.

Thus, the forces and displacement at the jth end and kth end in the local axes system are connected using the stiffness matrix as follows:

$$\begin{Bmatrix} P_t \\ P_r \\ P_v \\ P_h \\ P_s \\ P_w \end{Bmatrix} = \begin{bmatrix} k_{tt} & 0 & 0 & k_{th} & 0 & 0 \\ 0 & 0 & 0 & 0 & 0 & 0 \\ 0 & 0 & 0 & 0 & 0 & 0 \\ k_{ht} & 0 & 0 & k_{hh} & 0 & 0 \\ 0 & 0 & 0 & 0 & 0 & 0 \\ 0 & 0 & 0 & 0 & 0 & 0 \end{bmatrix} \begin{Bmatrix} \delta_t \\ \delta_r \\ \delta_v \\ \delta_h \\ \delta_s \\ \delta_w \end{Bmatrix}$$

$$\{P_T\}_i = [K_T]_i \{\delta_T\}_i \tag{5.1}$$

In the previous equation, 'T' represents the truss element and 'i' represents the member number. Now, the forces in the local axes system and the reference axes system are connected using the transformation matrix.

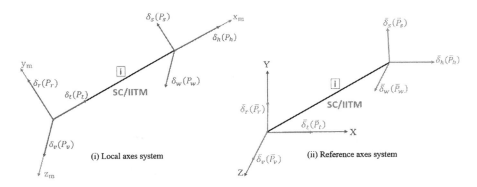

FIGURE 5.1 Local and reference axes system.

$$\begin{Bmatrix} P_{jx} \\ P_{jy} \\ P_{jz} \\ P_{kx} \\ P_{ky} \\ P_{kz} \end{Bmatrix} = \begin{bmatrix} C_{11} & C_{12} & C_{13} & 0 & 0 & 0 \\ C_{21} & C_{22} & C_{32} & 0 & 0 & 0 \\ C_{31} & C_{32} & C_{33} & 0 & 0 & 0 \\ 0 & 0 & 0 & C_{11} & C_{12} & C_{13} \\ 0 & 0 & 0 & C_{21} & C_{22} & C_{32} \\ 0 & 0 & 0 & C_{31} & C_{32} & C_{33} \end{bmatrix} \quad (5.2)$$

$$\{P_T\}_i = [T_T]_i \{\bar{P}_T\}_i$$

The transformation matrix for the truss member can be written as follows:

$$[T_T]_i = \begin{bmatrix} [C_T]_i & [0] \\ [0] & [C_T]_i \end{bmatrix}$$

Hence, the following equations will be valid for the truss member also.

$$\{\delta_T\}_i = [T_T]_i \{\bar{\delta}_T\}_i$$
$$[\bar{K}_T]_i = [T_T]_i^T [K_T]_i [T_T]_i \quad (5.3)$$

Further, we can also say,

$$\{p\}_i = [k_T]_i \{\delta_T\}_i + \{FP\}_i$$
$$\{\bar{p}\}_i = [\bar{k}_T]_i \{\bar{\delta}_T\}_i + \{\overline{FP}\}_i \quad (5.4)$$

For a truss member that is arbitrarily oriented in space, one can either use Y-Z-X transformation or Z-Y-X transformation. Therefore, all equations for both the transformations derived previously for the three-dimensional beam element for obtaining the ψ angle are applicable without any changes to the truss member. If the truss

Analysis of Special Members

members are loaded only at the joints (which is a common phenomenon), orientation of the local axes of the member with respect to the reference axes of the system is not important. In such a case, the local axes system can be positioned so that the ψ angle is practically zero.

Thus,

$$C_y = \begin{bmatrix} C_x & C_y & C_z \\ \dfrac{-C_xC_y}{\sqrt{C_x^2+C_z^2}} & \sqrt{C_x^2+C_z^2} & \dfrac{-C_yC_z}{\sqrt{C_x^2+C_z^2}} \\ -\dfrac{C_z}{\sqrt{C_x^2+C_z^2}} & 0 & \dfrac{C_x}{\sqrt{C_x^2+C_z^2}} \end{bmatrix}$$

$$C_z = \begin{bmatrix} C_x & C_y & C_z \\ \dfrac{-C_y}{\sqrt{C_x^2+C_y^2}} & \dfrac{C_x}{\sqrt{C_x^2+C_y^2}} & 0 \\ -\dfrac{C_xC_z}{\sqrt{C_x^2+C_y^2}} & \dfrac{-C_yC_z}{\sqrt{C_x^2+C_y^2}} & \sqrt{C_x^2+C_y^2} \end{bmatrix}$$

5.2 SPECIAL ELEMENTS

The structural members with a varying cross-section and non-uniform moment of inertia can be called special elements. Consider an element with one end fixed and another end hinged, as shown in Figure 5.2. The support conditions considered here are different from those of the standard beam element considered previously. The special member will then be converted to the conventional member with the procedure followed so far. The problem can now be converted into a fixed beam with the

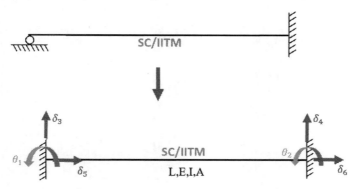

FIGURE 5.2 Degrees-of-freedom.

hinge introduced at one end, with a uniform moment of inertia, Young's modulus and area of cross-section. But, there is an unrestrained degree-of-freedom at the end with the hinge, the remaining degrees being restrained. The degrees-of-freedom are then marked for the beam element.

Thus,

Unrestrained degrees-of-freedom = 1 (θ_1)
Restrained degrees-of-freedom = 5 (θ_2, θ_3, δ_4, δ_5, δ_6)

For a standard beam element, the stiffness matrix can be written as follows:

$$[K]_i = \begin{bmatrix} \dfrac{4EI}{l} & \dfrac{2EI}{l} & \dfrac{6EI}{l^2} & -\dfrac{6EI}{l^2} & 0 & 0 \\ \dfrac{2EI}{l} & \dfrac{4EI}{l} & \dfrac{6EI}{l^2} & -\dfrac{6EI}{l^2} & 0 & 0 \\ \dfrac{6EI}{l^2} & \dfrac{6EI}{l^2} & \dfrac{12EI}{l^3} & -\dfrac{12EI}{l^3} & 0 & 0 \\ -\dfrac{6EI}{l^2} & -\dfrac{6EI}{l^2} & -\dfrac{12EI}{l^3} & \dfrac{12EI}{l^3} & 0 & 0 \\ 0 & 0 & 0 & 0 & \dfrac{AE}{l} & -\dfrac{AE}{l} \\ 0 & 0 & 0 & 0 & -\dfrac{AE}{l} & \dfrac{AE}{l} \end{bmatrix}$$

For this case, the stiffness matrix can be split into the following submatrices:

$$[K] = \begin{bmatrix} [k_{uu}]_{1\times1} & [k_{ur}]_{1\times5} \\ [k_{ru}]_{5\times1} & [k_{rr}]_{5\times5} \end{bmatrix}$$

where:

$$[k_{uu}] = \dfrac{4EI}{l}$$

$$[k_{rr}] = \begin{bmatrix} \dfrac{4EI}{l} & \dfrac{6EI}{l^2} & -\dfrac{6EI}{l^2} & 0 & 0 \\ \dfrac{6EI}{l^2} & \dfrac{12EI}{l^3} & -\dfrac{12EI}{l^3} & 0 & 0 \\ -\dfrac{6EI}{l^2} & -\dfrac{12EI}{l^3} & \dfrac{12EI}{l^3} & 0 & 0 \\ 0 & 0 & 0 & \dfrac{AE}{l} & -\dfrac{AE}{l} \\ 0 & 0 & 0 & -\dfrac{AE}{l} & \dfrac{AE}{l} \end{bmatrix}$$

Analysis of Special Members

$$[k_{ru}] = \begin{bmatrix} \dfrac{2EI}{l} \\ \dfrac{6EI}{l^2} \\ -\dfrac{6EI}{l^2} \\ 0 \\ 0 \end{bmatrix}$$

$$[k_{ru}] = \begin{bmatrix} \dfrac{2EI}{l} & \dfrac{6EI}{l^2} & -\dfrac{6EI}{l^2} & 0 & 0 \end{bmatrix}$$

Now, the stiffness matrix for the special element is given by,

$$[K]_{special} = [K_{rr}] - [K_{ru}][K_{uu}]^{-1}[K_{ur}] \tag{5.5}$$

By substituting the values of submatrices in the previous equation,

$$[k_{rr}] = \begin{bmatrix} \dfrac{3EI}{l} & \dfrac{3EI}{l^2} & -\dfrac{3EI}{l^2} & 0 & 0 \\ \dfrac{3EI}{l^2} & \dfrac{3EI}{l^3} & -\dfrac{3EI}{l^3} & 0 & 0 \\ -\dfrac{3EI}{l^2} & -\dfrac{3EI}{l^3} & \dfrac{3EI}{l^3} & 0 & 0 \\ 0 & 0 & 0 & \dfrac{AE}{l} & -\dfrac{AE}{l} \\ 0 & 0 & 0 & -\dfrac{AE}{l} & \dfrac{AE}{l} \end{bmatrix}$$

Thus, it can be seen that the stiffness method can be conveniently modified to analyze any element with varied boundary conditions or support conditions. It is very interesting to note that we are using the same procedure as developed for the conventional beam element to derive the stiffness matrix of a special element.

5.3 NON-PRISMATIC MEMBERS

Non-prismatic members are the common application in offshore structures. Depending upon the topside requirements, there may be a possibility that the beam moment of inertia can vary depending upon the span length. This kind of problem can be handled using the substructure technique. We already know that the beam element with special support conditions can be handled as a conventional problem by partitioning the matrices. This is called the substructure technique.

Example Problem with Computer Program

EXAMPLE 5.1:

Analyze the beam shown in Figure 5.3 using the substructure technique.

SOLUTION:

1. *Mark the degrees-of-freedom:*

   Assume a structural hinge at point 'B', and mark the unrestrained and restrained degrees-of-freedom, as shown in Figure 5.4.
   Unrestrained degrees-of-freedom = 3
   Restrained degrees-of-freedom = 6
   Thus, the total number of degrees-of-freedom is nine.

   Global labels:

   AB = [4,1,6,2,8,3]
   BC = [1,5,2,7,3,9]

2. *Stiffness matrix:*

   The conventional stiffness matrices for both the members can be written as follows:

$$[K_{AB}] = E \begin{array}{c} \\ \end{array} \begin{bmatrix} 0.0047 & 0.0023 & 0.0018 & -0.0018 & 0 & 0 \\ 0.0023 & 0.0047 & 0.0018 & -0.0018 & 0 & 0 \\ 0.0018 & 0.0018 & 0.0009 & 0.0009 & 0 & 0 \\ -0.0018 & -0.0018 & -0.0009 & 0.0009 & 0 & 0 \\ 0 & 0 & 0 & 0 & 0.0469 & -0.0469 \\ 0 & 0 & 0 & 0 & -0.0469 & 0.0469 \end{bmatrix} \begin{array}{c} 4 \\ 1 \\ 6 \\ 2 \\ 8 \\ 3 \end{array}$$

with column labels (4, 1, 6, 2, 8, 3)

$$[K_{BC}] = E \begin{bmatrix} 0.0062 & 0.0031 & 0.0047 & -0.0047 & 0 & 0 \\ 0.0031 & 0.0062 & 0.0047 & -0.0047 & 0 & 0 \\ 0.0047 & 0.0047 & 0.0047 & -0.0047 & 0 & 0 \\ -0.0047 & -0.0047 & -0.0047 & 0.0047 & 0 & 0 \\ 0 & 0 & 0 & 0 & 0.075 & -0.075 \\ 0 & 0 & 0 & 0 & -0.075 & 0.075 \end{bmatrix} \begin{array}{c} 1 \\ 5 \\ 2 \\ 7 \\ 3 \\ 9 \end{array}$$

with column labels (1, 5, 2, 7, 3, 9)

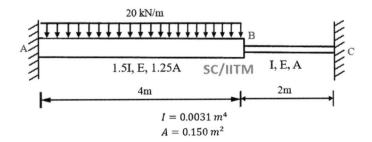

$I = 0.0031\ m^4$
$A = 0.150\ m^2$

FIGURE 5.3 Beam example.

Analysis of Special Members

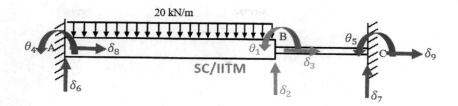

FIGURE 5.4 Unrestrained and restrained degrees-of-freedom.

The total stiffness matrix is developed by assembling the previous stiffness matrices.

$$[K_{total}] = E \begin{bmatrix} 0.0108 & 0.0029 & 0 & 0.0023 & 0.0031 & 0.0017 & -0.0046 & 0 & 0 \\ 0.0029 & 0.0055 & 0 & -0.0017 & 0.0046 & -0.0009 & -0.0046 & 0 & 0 \\ 0 & 0 & 0.1219 & 0 & 0 & 0 & 0 & -0.0469 & -0.0750 \\ 0.0023 & -0.0017 & 0 & 0.0046 & 0 & 0.0017 & 0 & 0 & 0 \\ 0.0031 & 0.0046 & 0 & 0 & 0.0062 & 0 & -0.0046 & 0 & 0 \\ 0.0017 & -0.0009 & 0 & 0.0017 & 0 & 0.0009 & 0 & 0 & 0 \\ -0.0046 & -0.0046 & 0 & 0 & -0.0046 & 0 & 0.0046 & 0 & 0 \\ 0 & 0 & -0.0469 & 0 & 0 & 0 & 0 & 0.0469 & 0 \\ 0 & 0 & -0.0750 & 0 & 0 & 0 & 0 & 0 & 0.0750 \end{bmatrix}$$

Now, the following submatrices can be written from the total stiffness matrix.

$$[K_{uu}] = E \begin{bmatrix} 0.0108 & 0.0029 & 0 \\ 0.0029 & 0.0055 & 0 \\ 0 & 0 & 0.1219 \end{bmatrix}$$

$$[K_{uu}]^{-1} = \frac{1}{E} \begin{bmatrix} 107.2916 & -56.4693 & 0 \\ -56.4693 & 210.8186 & 0 \\ 0 & 0 & 8.2051 \end{bmatrix}$$

$$[K_{rr}] = E \begin{bmatrix} 0.0046 & 0 & 0.0017 & 0 & 0 & 0 \\ 0 & 0.0062 & 0 & -0.0046 & 0 & 0 \\ 0.0017 & 0 & 0.0009 & 0 & 0 & 0 \\ 0 & -0.0046 & 0 & 0.0046 & 0 & 0 \\ 0 & 0 & 0 & 0 & 0.0469 & 0 \\ 0 & 0 & 0 & 0 & 0 & 0.0750 \end{bmatrix}$$

$$[K_{ur}] = E \begin{bmatrix} 0.0023 & 0.0031 & 0.0017 & -0.0046 & 0 & 0 \\ -0.0017 & 0.0046 & -0.0009 & -0.0046 & 0 & 0 \\ 0 & 0 & 0 & 0 & -0.0469 & -0.0750 \end{bmatrix}$$

$$[K_{ru}] = E \begin{bmatrix} 0.0023 & -0.0017 & 0 \\ 0.0031 & 0.0046 & 0 \\ 0.0017 & -0.009 & 0 \\ -0.0046 & -0.0046 & 0 \\ 0 & 0 & -0.0469 \\ 0 & 0 & -0.0750 \end{bmatrix} \begin{matrix} ④ \\ ⑤ \\ ⑥ \\ ⑦ \\ ⑧ \\ ⑨ \end{matrix}$$

with columns ①, ②, ③.

$$[K]_{special} = [K_{rr}] - [K_{ru}][K_{uu}]^{-1}[K_{ur}]$$

$$[K]_{special} = E \begin{bmatrix} 0.0030 & 0.0012 & 0.0007 & -0.007 & 0 & 0 \\ 0.0012 & 0.0022 & 0.0006 & -0.006 & 0 & 0 \\ 0.0007 & 0.0004 & 0.0002 & -0.0002 & 0 & 0 \\ -0.0007 & -0.0004 & -0.0002 & 0.0002 & 0 & 0 \\ 0 & 0 & 0 & 0 & 0.0288 & -0.0288 \\ 0 & 0 & 0 & 0 & -0.0288 & 0.0288 \end{bmatrix}$$

3. *Calculation of fixed end moments and joint load vector:*

$$[FEM_{AB}] = \begin{Bmatrix} 26.667 \\ -26.667 \\ 40 \\ 40 \\ 0 \\ 0 \end{Bmatrix}, \quad [FEM_{BC}] = [0]$$

$$\{J_L\} = \begin{Bmatrix} 26.667 \\ -40 \\ 0 \\ -26.667 \\ 0 \\ -40 \\ 0 \\ 0 \\ 0 \end{Bmatrix}_{9 \times 1} \begin{matrix} ① \\ ② \\ ③ \\ ④ \\ ⑤ \\ ⑥ \\ ⑦ \\ ⑧ \\ ⑨ \end{matrix}$$

4. *Calculation of end moments and reactions:*

$$\{R_r\} = [K_{ru}][K_{uu}]^{-1}\{J_{Lu}\} - \{J_{Lr}\}$$

Analysis of Special Members

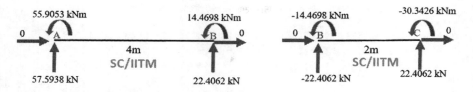

FIGURE 5.5 Member end forces and moments.

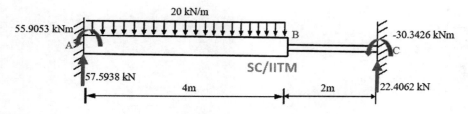

FIGURE 5.6 Final end forces and moments.

$$\{M_1\} = \begin{Bmatrix} 55.9053 \\ 14.4698 \\ 57.5938 \\ 22.4062 \\ 0 \\ 0 \end{Bmatrix}, \quad \{M_2\} = \begin{Bmatrix} -14.4698 \\ -30.3426 \\ -22.4062 \\ -22.4062 \\ 0 \\ 0 \end{Bmatrix}$$

The member end forces and moments are shown in Figure 5.5 and the fixed end moments and reactions are shown in Figure 5.6.

MATLAB® program

```
%% stiffness matrix method for non prismatic members
% Input
clc;
clear;
n = 2; % number of members
I1 = 0.0031; %value in m4
A1 = 0.15; %value in m2
I = [1.5*I1 I1]; %Moment of inertis in m4
L = [4 2]; % length in m
A = [1.25*A1 A1]; % Area in m2
uu = 3; % Number of unrestrained degrees of freedom
ur = 6; % Number of restrained degrees of freedom
uul = [1 2 3]; % global labels of unrestrained dof
url = [4 5 6 7 8 9]; % global labels of restrained dof
l1 = [4 1 6 2 8 3]; % Global labels for member 1
```

```
l2 = [1 5 2 7 3 9]; % Global labels for member 2
l= [l1; l2];
dof = uu+ur;
Ktotal = zeros (dof);
fem1= [26.67 -26.67 40 40 0 0]; % Local Fixed end moments of
member 1
fem2= [0 0 0 0 0 0]; % Local Fixed end moments of member 2

%% rotation coefficients for each member
rc1 = 4.*I./L;
rc2 = 2.*I./L;
rc3 = A./L;

%% stiffness matrix 4 by 4 (axial deformation neglected)
for i = 1:n
    Knew = zeros (dof);
    k1 = [rc1(i); rc2(i); (rc1(i)+rc2(i))/L(i);
(-(rc1(i)+rc2(i))/L(i)); 0; 0];
    k2 = [rc2(i); rc1(i); (rc1(i)+rc2(i))/L(i);
(-(rc1(i)+rc2(i))/L(i)); 0; 0];
    k3 = [(rc1(i)+rc2(i))/L(i); (rc1(i)+rc2(i))/L(i);
(2*(rc1(i)+rc2(i))/(L(i)^2)); (-2*(rc1(i)+rc2(i))/(L(i)^2));
0; 0];
    k4 = -k3;
    k5 = [0; 0; 0; 0; rc3(i); -rc3(i)];
    k6 = [0; 0; 0; 0; -rc3(i); rc3(i)];
    K = [k1 k2 k3 k4 k5 k6];
    fprintf ('Member Number =');
    disp (i);
    fprintf ('Local Stiffness matrix of member, [K] = \n');
    disp (K);
    for p = 1:6
        for q = 1:6
            Knew((l(i,p)),(l(i,q))) =K(p,q);
        end
    end
    Ktotal = Ktotal + Knew;
    if i == 1
        Kg1=K;
    elseif i == 2
        Kg2 =K;
    end
end
fprintf ('Stiffness Matrix of complete structure, [Ktotal] = \n');
disp (Ktotal);

%% Kuu matrix
Kuu = zeros(uu);
for x=1:uu
    for y=1:uu
        Kuu(x,y)= Ktotal(x,y);
```

Analysis of Special Members

```
            end
    end
fprintf ('Unrestrained Stiffness sub-matrix, [Kuu] = \n');
disp (Kuu);
KuuInv= inv(Kuu);
fprintf ('Inverse of Unrestrained Stiffness sub-matrix,
[KuuInverse] = \n');
disp (KuuInv);

%% Krr matrix
Krr = zeros(ur);
for x=(1+uu):dof
    for y=(1+uu):dof
        Krr((x-uu),(y-uu))= Ktotal(x,y);
    end
end
fprintf ('Restrained Stiffness sub-matrix, [Krr] = \n');
disp (Krr);

%% Kur matrix
Kur = zeros(uu,ur);
for x=1:uu
    for y=(1+uu):dof
        Kur((x),(y-uu))= Ktotal(x,y);
    end
end
fprintf ('[Kur] = \n');
disp (Kur);

%% Kru matrix
Kru = zeros(ur,uu);
for x=(1+uu):dof
    for y=1:uu
        Kru((x-uu),(y))= Ktotal(x,y);
    end
end
fprintf ('[Kru] = \n');
disp (Kru);

%% K modified
Kmod = Krr - (Kru*KuuInv*Kur);
fprintf ('Modified Stiffness matrix = \n');
disp (Kmod);

%% Creation of joint load vector
jl= [26.67; -40; 0; -26.67; 0; -40; 0; 0; 0]; % values given
in kN or kNm
jlu = [26.67; -40; 0]; % load vector in unrestrained dof
jlr = [-26.67; 0; -40; 0; 0; 0]; % load vector in restrained
dof
delu = KuuInv*jlu;
```

```
fprintf ('Joint Load vector, [Jl] = \n');
disp (jl');
fprintf ('Unrestrained displacements, [DelU] = \n');
disp (delu');
Rr = (Kru*KuuInv*jlu) - jlr;
fprintf ('Rr vector = \n');
disp (Rr');
delr = [0; 0; 0; 0; 0; 0];
del = [delu; delr];
deli= zeros (6,1);
for i = 1:n
    for p = 1:6
        deli(p,1) = del((l(i,p)),1) ;
    end
    if i == 1
        delbar1 = deli;
        mbar1= (Kg1 * delbar1)+fem1';
        fprintf ('Member Number =');
        disp (i);
        fprintf ('Global displacement matrix [DeltaBar] = \n');
        disp (delbar1');
        fprintf ('Global End moment matrix [MBar] = \n');
        disp (mbar1');
    elseif i == 2
        delbar2 = deli;
        mbar2= (Kg2 * delbar2)+fem2';
        fprintf ('Member Number =');
        disp (i);
        fprintf ('Global displacement matrix [DeltaBar] = \n');
        disp (delbar2');
        fprintf ('Global End moment matrix [MBar] = \n');
        disp (mbar2');
    end
end

%% check
mbar = [mbar1'; mbar2'];
jf = zeros(dof,1);
for a=1:n
    for b=1:4 % size of k matrix
        d = l(a,b);
        jfnew = zeros(dof,1);
        jfnew(d,1)=mbar(a,b);
        jf=jf+jfnew;
    end
end
fprintf ('Joint forces = \n');
disp (jf');
```

Analysis of Special Members 217

MATLAB output:

```
Member Number =    1
Local Stiffness matrix of member, [K] =

    0.0046    0.0023    0.0017   -0.0017         0         0
    0.0023    0.0046    0.0017   -0.0017         0         0
    0.0017    0.0017    0.0009   -0.0009         0         0
   -0.0017   -0.0017   -0.0009    0.0009         0         0
         0         0         0         0    0.0469   -0.0469
         0         0         0         0   -0.0469    0.0469

Member Number =    2
Local Stiffness matrix of member, [K] =

    0.0062    0.0031    0.0046   -0.0046         0         0
    0.0031    0.0062    0.0046   -0.0046         0         0
    0.0046    0.0046    0.0046   -0.0046         0         0
   -0.0046   -0.0046   -0.0046    0.0046         0         0
         0         0         0         0    0.0750   -0.0750
         0         0         0         0   -0.0750    0.0750

Stiffness Matrix of complete structure, [Ktotal] =

 0.0108  0.0029       0   0.0023  0.0031  0.0017 -0.0046       0        0
 0.0029  0.0055       0  -0.0017  0.0046 -0.0009 -0.0046       0        0
      0       0  0.1219        0       0       0       0 -0.0469  -0.0750
 0.0023 -0.0017       0   0.0046       0  0.0017       0       0        0
 0.0031  0.0046       0        0  0.0062       0 -0.0046       0        0
 0.0017 -0.0009       0   0.0017       0  0.0009       0       0        0
-0.0046 -0.0046       0        0 -0.0046       0  0.0046       0        0
      0       0 -0.0469        0       0       0       0  0.0469        0
      0       0 -0.0750        0       0       0       0       0   0.0750

Unrestrained Stiffness sub-matrix, [Kuu] =

    0.0108    0.0029         0
    0.0029    0.0055         0
         0         0    0.1219

Inverse of Unrestrained Stiffness sub-matrix, [KuuInverse] =

  107.2916   -56.4693         0
  -56.4693   210.8186         0
         0         0    8.2051
```

Restrained Stiffness sub-matrix, [Krr] =

```
    0.0046         0    0.0017         0         0         0
         0    0.0062         0   -0.0046         0         0
    0.0017         0    0.0009         0         0         0
         0   -0.0046         0    0.0046         0         0
         0         0         0         0    0.0469         0
         0         0         0         0         0    0.0750
```

[Kur] =

```
    0.0023    0.0031    0.0017   -0.0046         0         0
   -0.0017    0.0046   -0.0009   -0.0046         0         0
         0         0         0         0   -0.0469   -0.0750
```

[Kru] =

```
    0.0023   -0.0017         0
    0.0031    0.0046         0
    0.0017   -0.0009         0
   -0.0046   -0.0046         0
         0         0   -0.0469
         0         0   -0.0750
```

Modified Stiffness matrix =

```
    0.0030    0.0012    0.0007   -0.0007         0         0
    0.0012    0.0022    0.0006   -0.0006         0         0
    0.0007    0.0006    0.0002   -0.0002         0         0
   -0.0007   -0.0006   -0.0002    0.0002         0         0
         0         0         0         0    0.0288   -0.0288
         0         0         0         0   -0.0288    0.0288
```

Joint Load vector, [Jl] =

```
   26.6700  -40.0000         0  -26.6700         0  -40.0000         0         0         0
```

Unrestrained displacements, [DelU] =

```
   1.0e+03 *
    5.1202   -9.9388         0
```

Rr vector =

```
   55.9053  -30.3426   57.5938   22.4062         0         0
```

Analysis of Special Members

```
Member Number =        1
Global displacement matrix [DeltaBar] =

   1.0e+03 *
        0    5.1202    0   -9.9388    0    0

Global End moment matrix [MBar] =

   55.9053   14.4698   57.5938   22.4062    0    0

Member Number =        2
Global displacement matrix [DeltaBar] =

   1.0e+03 *
    5.1202       0   -9.9388    0    0    0

Global End moment matrix [MBar] =

  -14.4698  -30.3426  -22.4062   22.4062    0    0

Joint forces =

    0   -0.0000    0   55.9053  -30.3426   57.5938   22.4062    0    0
```

Appendix

STIFFNESS MATRIX DERIVATION USING THE ENERGY METHOD

The energy method is the basis for the derivation of the member stiffness matrix. This method can be applied to analyze structures with any geometric shape. The major assumption made in this method is that the members does not have any geometric non-linearity or P-Δ effect. Based on this assumption, the following assumption is valid:

$$\frac{d^2y}{dx^2} = \frac{M}{EI} \tag{A.1}$$

where,
- y is the displacement,
- M is the moment and
- EI is the flexural rigidity.

The same equation can be extended to standard potential energy. The principle of stationary potential energy states that, "When a system is in a state of equilibrium, the first derivative of the local potential energy of the structural system with respect to the joint displacement is zero." Mathematically this can be expressed as,

$$\frac{\partial V}{\partial dj} = 0, \quad \text{for } j = 1, 2, \ldots n \tag{A.2}$$

where,
- V is the total potential energy,
- dj is the joint displacement and
- j is the number of joints in the structure.

The total potential energy has two components such as external potential energy and internal potential energy. It is given by,

$$V = W_{\text{ext}} + W_{\text{int}} \tag{A.3}$$

The external potential energy is the sum of the products of applied loads and corresponding displacements in the structure. Thus,

$$W_{\text{ext}} = \sum_{j=1}^{n} P_j \delta_j \tag{A.4}$$

where,

P_j is the applied load at point j and
δ_j is the displacement at point j in the direction of P_j.

The internal potential energy is the sum of the products of internal stresses and strains. Mathematically,

$$W_{int} = \int_{vol} \int_0^\varepsilon \sigma * d\varepsilon * dv \tag{A.5}$$

where,

σ is the stress at any internal point,
$d\varepsilon$ is the strain at the same point and
$\int_0^\varepsilon \sigma * d\varepsilon$ represents the amount of internal energy created at any point in a system that has unit volume.

To apply this equation for the analysis of the structure, the volume of the structure should be known, which is very hard to find for non-rectilinear members. Thus, a different method or analogy is used for solving the problem.

Alternatively, internal potential energy is also equivalent to the internal work done. Internal work performed on a system stores energy in the system. This is commonly known as **strain energy**, U. Internal potential energy and strain energy are related as follows:

$$W_{int} = -U \tag{A.6}$$

Substituting equation A.6 in equation A.3,

$$V = W_{ext} - U \tag{A.7}$$

Substituting equation A.7 in equation A.2,

$$\frac{\partial}{\partial dj}(W_{ext} - U) = 0$$

or

$$\frac{\partial}{\partial dj}(U - W_{ext}) = 0 \tag{A.8}$$

This equation is used to find the elements of the stiffness matrix.

A.1 AXIAL STRAIN ENERGY

Axial strain energy is the primary type of strain energy stored in members under axial forces like truss members. Consider a truss member of length '*l*' and cross-sectional

Appendix

area 'A' with nodes j and k. The member is oriented along the X-axis, as shown in Figure A.1.

The member is displaced along the direction of the X-axis with displacements δ_t and δ_h at nodes j and k respectively. The displaced position of the truss member is shown in Figure A.1.

Axial strain energy stored in the member is given by,

$$U_{axial} = \int_0^L \frac{N^2}{2} \frac{1}{AE} dx \qquad (A.9)$$

Considering uniform AE throughout the section,

$$U_{axial} = \frac{N^2}{2AE} L \qquad (A.10)$$

By considering the axial displacements at both the nodes, the net elongation is given by,

$$e = \delta_h - \delta_t \qquad (A.11)$$

For uniform extension of the member under the axial load, the net elongation is given by,

$$e = \frac{NL}{AE} \qquad (A.12a)$$

$$N = e\left(\frac{AE}{L}\right) \qquad (A.12b)$$

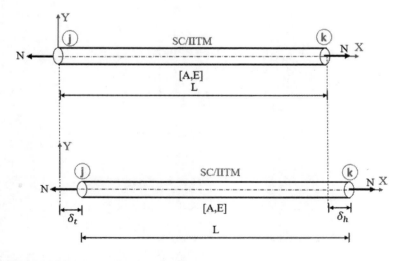

FIGURE A.1 Original and displaced position of member.

Substituting equation A.12b in equation A.10, the axial strain energy for uniform cross-section is given by,

$$U_{axial} = \left[e\left(\frac{AE}{L}\right) \right]^2 \frac{L}{2AE}$$

$$U_{axial} = \frac{e^2}{2L} AE \tag{A.13}$$

Substituting equation A.11 in equation A.13,

$$U_{axial} = \frac{AE}{2L}\left(\delta_h - \delta_t\right)^2 \tag{A.13a}$$

A.2 BENDING STRAIN ENERGY

Bending strain energy is the primary type of energy stored in beams and frames. Bending strain energy is given by,

$$U_{bending} = \int_O^L \frac{M^2 dx}{2EI} \tag{A.14}$$

where,

M is the moment causing bending on the ith member, as shown in Figure A.2.

The elastic curve is related to the loading diagram or the applied moment by the following relationship:

$$EI \frac{d^2 y}{dx^2} = M \tag{A.15}$$

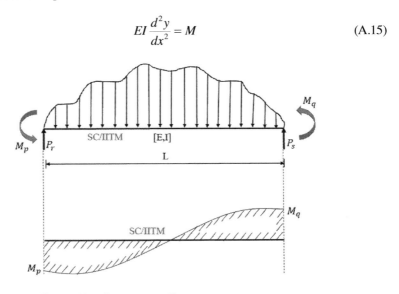

FIGURE A.2 Loading and bending moment diagram.

Appendix

Substituting equation A.15 in equation A.14, we get

$$U_{bending} = \int_0^L \left[EI \frac{d^2y}{dx^2} \right]^2 \frac{1}{2EI} dx$$

$$= \frac{1}{2EI}(EI)^2 \int_0^L \left[\frac{d^2y}{dx^2} \right]^2 dx \qquad (A.16)$$

$$= \frac{EI}{2} \int_0^L \left[\frac{d^2y}{dx^2} \right]^2 dx$$

The previous equation is valid for members with non-varying EI. Now, the previous equation should be expressed in terms of displacements mentioned in the deflected profile in Figure A.3.

Consider a beam of length 'L' with a uniformly distributed load of 'w' for the entire length. The member is oriented along the X-axis. M_p is the moment at the jth end, M_q is the moment at the kth end, P_r is the reaction at the jth end and P_s is the reaction at the kth end. The corresponding displacements at j and k ends are shown in the elastic curve.

Consider a section X-X at a distance 'x' from the jth end of the member. The moment at the section is taken as 'M'. Taking the moment about X-X from the left of the section,

$$M_p - \left(\frac{wx^2}{2} \right) - P_r(x) + M = 0$$

Rearranging the previous equation,

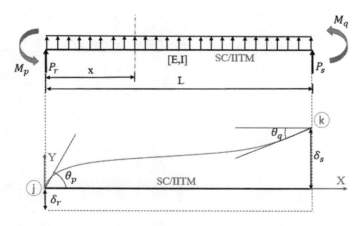

FIGURE A.3 Loading diagram and elastic curve.

$$M + M_p - P_r(x) - \left(\frac{wx^2}{2}\right) = 0 \qquad \text{(A.17)}$$

$$M = -M_p + P_r(x) + \left(\frac{wx^2}{2}\right) \qquad \text{(A.17a)}$$

Substituting equation A.17a in equation A.15,

$$EI\left(\frac{d^2y}{dx^2}\right) = -M_p + P_r(x) + \left(\frac{wx^2}{2}\right)$$

$$\left(\frac{d^2y}{dx^2}\right) = -\frac{M_p}{EI} + \frac{P_r(x)}{EI} + \left(\frac{wx^2}{2EI}\right) \qquad \text{(A.17b)}$$

By integrating the previous equation once, we will get the slope equation.

$$\left(\frac{dy}{dx}\right) = -\frac{M_p}{EI}x + \frac{P_r(x^2)}{2EI} + \left(\frac{wx^3}{6EI}\right) + C_1 \qquad \text{(A.18)}$$

The following boundary conditions are applied in the previous equation to get the value of the constant.

1. $\dfrac{dy}{dx} = \theta_p$ at $x=0$.
2. $y = \delta_r$ at $x=0$.

Substituting $x=0$ in equation A.18,

$$C_1 = \theta_p$$

By substituting the value of the constant in equation A.18, the slope equation is given by,

$$\left(\frac{dy}{dx}\right) = -\frac{M_p}{EI}x + \frac{P_r(x^2)}{2EI} + \left(\frac{wx^3}{6EI}\right) + \theta_p \qquad \text{(A.19)}$$

The previous equation is again integrated to get the equation for displacement,

$$y = -\frac{M_p}{EI}x^2 + \frac{P_r(x^3)}{6EI} + \left(\frac{wx^4}{24EI}\right) + \theta_p x + C_2$$

Again applying the boundary conditions to get the value of the constant in the displacement equation,

$$C_2 = \delta_r$$

Appendix

Hence,

$$y = \delta_r + \theta_p x - \frac{M_p}{2EI}x^2 + \frac{P_r(x^3)}{6EI} + \left(\frac{wx^4}{24EI}\right) \quad \text{(A.20)}$$

Now, the following set of boundary conditions are applied in equations A.19 and A.20,

1. $\dfrac{dy}{dx} = \theta_q$ at $x=L$.
2. $y = \delta_s$ at $x=L$.

Equation A.19 becomes,

$$\theta_q = \theta_p - \frac{M_p}{EI}L + \frac{P_r(L^2)}{2EI} + \left(\frac{wL^3}{6EI}\right) \quad \text{(A.19a)}$$

Equation A.20 becomes,

$$\delta_s = \delta_r + \theta_p L - \frac{M_p}{2EI}L^2 + \frac{P_r(L^3)}{6EI} + \left(\frac{wL^4}{24EI}\right) \quad \text{(A.20a)}$$

Rewriting equations A.19a and A.20a as follows:

$$\frac{L}{EI}\left[M_p - \frac{P_r L}{2}\right] = \theta_p - \theta_q + \left(\frac{wL^3}{6EI}\right)$$

$$\frac{L^2}{2EI}\left[M_p - \frac{P_r L}{2}\right] = \delta_r - \delta_s + \theta_p L + \left(\frac{wL^4}{24EI}\right)$$

The previous equations are solved to get M_p and P_r as follows:

$$M_p = \frac{EI}{L}\left[4\theta_p + 2\theta_q + \frac{6\delta_r}{L} - \frac{6\delta_s}{L}\right] - \left(\frac{wL^2}{12}\right) \quad \text{(A.21)}$$

$$P_r = \frac{6EI}{L^2}\left[\theta_p + \theta_q + \frac{2\delta_r}{L} - \frac{2\delta_s}{L}\right] - \left(\frac{wL}{2}\right) \quad \text{(A.22)}$$

Equations A.21 and A.22 give the stiffness coefficients of the first and third column of the stiffness matrix, respectively.

Thus at jth end of the member,

$$M_p = \frac{EI}{L}\left[4\theta_p + 2\theta_q + \frac{6\delta_r}{L} - \frac{6\delta_s}{L}\right] - \left(\frac{wL^2}{12}\right) \quad \text{(A.23)}$$

$$P_r = \frac{6EI}{L^2}\left[\theta_p + \theta_q + \frac{2\delta_r}{L} - \frac{2\delta_s}{L}\right] - \left(\frac{wL}{2}\right) \quad (A.24)$$

Similarly, at kth end,

$$M_q = \frac{EI}{L}\left[2\theta_p + 4\theta_q + \frac{6\delta_r}{L} - \frac{6\delta_s}{L}\right] - \left(\frac{wL^2}{12}\right) \quad (A.25)$$

$$P_s = \frac{6EI}{L^2}\left[-\theta_p - \theta_q - \frac{2\delta_r}{L} + \frac{2\delta_s}{L}\right] - \left(\frac{wL}{2}\right) \quad (A.26)$$

In the previous equations, $\dfrac{wL^2}{12}$ is the fixed end moment of the member with udl under consideration and $wL/2$ is the reaction.

Substituting equations A.23 and A.24 in equation A.20, we get

$$\begin{aligned} y = \delta_r + \theta_p x &- \frac{x^2}{2EI}\left[\frac{EI}{L}\left[4\theta_p + 2\theta_q + \frac{6\delta_r}{L} - \frac{6\delta_s}{L}\right] - \text{FEM}_p\right] \\ &+ \frac{x^3}{6EI}\left[\frac{6EI}{L^2}\left[-\theta_p - \theta_q - \frac{2\delta_r}{L} + \frac{2\delta_s}{L}\right] - \left(\frac{wL}{2}\right)\right] + \left(\frac{wx^4}{24EI}\right) \end{aligned} \quad (A.27)$$

The previously mentioned is rearranged in such a way by retaining the displacement parameters and replacing the remaining terms with an arbitrary function $G(x)$. $G(x)$ is not a function of displacement and hence, on differentiation with respect to any displacement, will become zero.

Thus, equation A.27 becomes

$$y = \delta_r + \theta_p x - \frac{x^2}{L}\left[2\theta_p + \theta_q + \frac{3\delta_r}{L} - \frac{3\delta_s}{L}\right] + \frac{x^3}{L^2}\left[\theta_p + \theta_q + \frac{2\delta_r}{L} - \frac{2\delta_s}{L}\right] + G(x)$$

On differentiation,

$$y' = \theta_p - \frac{2x}{L}\left[2\theta_p + \theta_q + \frac{3\delta_r}{L} - \frac{3\delta_s}{L}\right] + \frac{3x^2}{L^2}\left[\theta_p + \theta_q + \frac{2\delta_r}{L} - \frac{2\delta_s}{L}\right] + G'(x) \quad (A.28)$$

$$y'' = -\frac{2}{L}\left[2\theta_p + \theta_q + \frac{3\delta_r}{L} - \frac{3\delta_s}{L}\right] + \frac{6x}{L^2}\left[\theta_p + \theta_q + \frac{2\delta_r}{L} - \frac{2\delta_s}{L}\right] + G''(x) \quad (A.29)$$

The square of the previously mentioned term should be used in the equation for bending to get the bending strain energy. This process will be cumbersome and hence the following analogy is followed. The displacement is separated into two terms as mentioned subsequently:

$$y = \bar{y} + y_f$$

Appendix

where,

$$\bar{y} = \delta_r + \theta_p x - \frac{x^2}{L}\left[2\theta_p + \theta_q + \frac{3\delta_r}{L} - \frac{3\delta_s}{L}\right] + \frac{x^3}{L^2}\left[\theta_p + \theta_q + \frac{2\delta_r}{L} - \frac{2\delta_s}{L}\right] \quad \text{(A.30)}$$

$$y_f = G(x)$$

Thus,

$$y' = \bar{y}' + y'_f$$

$$y'' = \bar{y}'' + y''_f \quad \text{(A.31)}$$

Substituting equation A.31 in equation A.16,

$$U_{\text{bending}} = \frac{EI}{2}\int_0^L \left[\bar{y}'' + y''_f\right]^2 dx$$

$$= \frac{EI}{2}\int_0^L \left[\left(\bar{y}''\right)^2 + \left(y''_f\right)^2 + 2\bar{y}''y''_f\right] dx$$

$$= \frac{EI}{2}\int_0^L \left[I_1 + I_2 + I_3\right]^2 dx$$

where,
I_1 is the function of displacement. The integral of functions I_2 and I_3 will become zero.

Hence, the bending strain energy can be written as follows:

$$U_{\text{bending}} = \frac{EI}{2}\int_0^L \left(\bar{y}''\right)^2 dx + C \quad \text{(A.32)}$$

Now, the previous equation is a function of displacement only. From equation A.29,

$$\bar{y}'' = -\frac{2}{L}\left[2\theta_p + \theta_q + \frac{3\delta_r}{L} - \frac{3\delta_s}{L}\right] + \frac{6x}{L^2}\left[\theta_p + \theta_q + \frac{2\delta_r}{L} - \frac{2\delta_s}{L}\right]$$

Substituting the previous equation in the bending strain energy equation A.32, we will get,

$$U_{\text{bending}} = \frac{2EI}{L}\left[\theta_p^2 + \theta_p\theta_q + \theta_p^2\right] + \frac{3}{L}\left[\left(\theta_p + \theta_q\right)\left(\delta_r - \delta_s\right)\right] + \frac{3}{L^2}\left(\delta_r - \delta_s\right)^2 + C \quad \text{(A.33)}$$

The previous equation is used for developing the stiffness matrix of a member.

For the member, the external work done is given by,

$$W_{ext} = \sum_{j=1}^{n} P_j \delta_i$$

Thus,

$$W_{ext} = M_p \theta_p + M_q \theta_q + P_r \delta_r + P_s \delta_s \qquad (A.34)$$

From equation A.7, the stationary potential energy is given by,

$$V = W_{ext} - U$$

The differential of the stationary potential energy equation will give a series of equations from which the stiffness matrix of the member can be developed.
The equations are as follows:

$$\frac{\partial V}{\partial \theta_p} = \frac{2EI}{L}\left[(2\theta_p + \theta_q) + \frac{3}{L}(\delta_r - \delta_s)\right] - M_p = 0$$

$$\frac{\partial V}{\partial \theta_q} = \frac{2EI}{L}\left[(\theta_p + 2\theta_q) + \frac{3}{L}(\delta_r - \delta_s)\right] - M_q = 0$$

$$\frac{\partial V}{\partial \delta_r} = \frac{2EI}{L}\left[\frac{3}{L}(\theta_p + \theta_q) + \frac{6}{L^2}(\delta_r - \delta_s)\right] - P_r = 0$$

$$\frac{\partial V}{\partial \delta_s} = \frac{2EI}{L}\left[-\frac{3}{L}(\theta_p + \theta_q) - \frac{6}{L^2}(\delta_r - \delta_s)\right] - P_s = 0$$

Writing the previous equations in matrix form,

$$\frac{EI}{L}\begin{bmatrix} 4 & 2 & \dfrac{6}{L} & -\dfrac{6}{L} \\ 2 & 4 & \dfrac{6}{L} & -\dfrac{6}{L} \\ \dfrac{6}{L} & \dfrac{6}{L} & \dfrac{12}{L^2} & -\dfrac{12}{L^2} \\ -\dfrac{6}{L} & -\dfrac{6}{L} & -\dfrac{12}{L^2} & \dfrac{12}{L^2} \end{bmatrix} \begin{Bmatrix} \theta_p \\ \theta_q \\ \delta_r \\ \delta_s \end{Bmatrix} = \begin{Bmatrix} M_p \\ M_q \\ P_r \\ P_s \end{Bmatrix}$$

Appendix

Thus, from the previous matrix equation, the stiffness matrix of the member can be written as follows:

$$k = \begin{bmatrix} \dfrac{4EI}{L} & \dfrac{2EI}{L} & \dfrac{6EI}{L^2} & -\dfrac{6EI}{L^2} \\ \dfrac{2EI}{L} & \dfrac{4EI}{L} & \dfrac{6EI}{L^2} & -\dfrac{6EI}{L^2} \\ \dfrac{6EI}{L^2} & \dfrac{6EI}{L^2} & \dfrac{12EI}{L^3} & -\dfrac{12EI}{L^3} \\ -\dfrac{6EI}{L^2} & -\dfrac{6EI}{L^2} & -\dfrac{12EI}{L^3} & \dfrac{12EI}{L^3} \end{bmatrix}$$

Bibliography

Beaufait, F.W., Rowan, W.H., Jr., Hoadley, P.G., Hackett, R.M. 1970. *Computer Methods of Structural Analysis*. New Jersey: Prentice Hall.

Chandrasekaran, S. 2013. *Advanced Marine Structures*. Video course on NPTEL portal. Available at: http://nptel.ac.in/courses/114106037/.

Chandrasekaran, S. 2013. *Dynamics of Ocean Structures*. Video Course on NPTEL portal. Available at: http://nptel.ac.in/courses/114106036/.

Chandrasekaran, S. 2013. *Health, Safety and Environmental Management (HSE) for Oil and Gas Industries*. Video course on NPTEL portal. Available at: http://nptel.ac.in/courses/114106017/.

Chandrasekaran, S. 2013. *Ocean Structures and Materials*. Video course on NPTEL portal. Available at: http://nptel.ac.in/courses/114106035/.

Chandrasekaran, S. 2014. *Advanced Theory on Offshore Plant FEED Engineering*, pp. 237. Changwon, South Korea: Changwon National University Press. ISBN: 978-89-969792-8-9.

Chandrasekaran, S. 2015. *Advanced Marine Structures*. Boca Raton, FL: CRC Press. ISBN: 9781498739689.

Chandrasekaran, S. 2015. *Dynamic Analysis and Design of Ocean Structures*. India: Springer. ISBN: 978-81-322-2276-7.

Chandrasekaran, S. 2015. *Dynamic Analysis of Offshore Structures*. Video course on MOOC, NPTEL portal at: https://onlinecourses.nptel.ac.in/noc15_oe01/preview.

Chandrasekaran, S. 2015. *HSE for Offshore and Petroleum Engineering*. Video course on MOOC, NPTEL portal at: https://onlinecourses.nptel.ac.in/noc15_oe02.

Chandrasekaran, S. 2016. *Risk and Reliability of Offshore Structures*. Video course on MOOC, NPTEL portal at: https://onlinecourses.nptel.ac.in/noc16_oe01.

Chandrasekaran, S. 2016. *Safety Practices for Offshore and Petroleum Engineers*. Video course on MOOC, NPTEL portal at: https://onlinecourses.nptel.ac.in/noc16_oe02.

Chandrasekaran, S. 2017. *Dynamic Analysis and Design of Ocean Structures*, 2nd Ed. Singapore: Springer. ISBN: 978-981-10-6088-5.

Chandrasekaran, S. 2017. *Dynamics of Offshore Structures*. Re-run video course.

Chandrasekaran, S. 2017. *Offshore Structures under Special Loads Including Fire Resistance*. Video course under MOOC, NPTEL portal http://nptel.ac.in/courses/114106043.

Chandrasekaran, S. 2018. *Computer Methods of Analysis of Offshore Structures*. Video course under MOOC, NPTEL portal http://nptel.ac.in/courses/114106045.

Chandrasekaran, S. 2018. *Risk and Reliability of Offshore Structures*. Video course re-run under MOOC.

Chandrasekaran, S., Bhattacharyya, S.K. 2012. *Analysis and Design of Offshore Structures with Illustrated Examples*. Changwon, South Korea: Human Resource Development Center for Offshore and Plant Engineering (HOPE Center), Changwon National University Press, pp. 285. ISBN: 978-89-963915-5-5.

Chandrasekaran, S., Kiran, P.A. 2018. Mathieu stability of offshore triceratops under postulated failure. *Ships and Offshore Structures* 13(2):143–148. DOI: 10.1080/17445302.2017.133578.

Chandrasekaran, S., Lognath, R.S. 2015. Dynamic analyses of Buoyant Leg Storage Regasification Platform (BLSRP) under regular waves: Experimental investigations. *Ships and Offshore Structures* 12(2):227–232.

Chandrasekaran, S., Lognath, R.S. 2017. Dynamic analyses of buoyant leg storage and regasification platforms: Numerical studies. *Journal of Marine Systems and Ocean Technology* 12(2):39–48. DOI: 10.1007/s40868-017-0022-6.

Chandrasekaran, S., Madhuri, S. 2015. Dynamic response of offshore triceratops: Numerical and experimental investigations. *Ocean Engineering* 109(15):401–409.

Chandrasekaran, S., Mayank, S. 2017. Dynamic analyses of stiffened triceratops under regular waves: Experimental investigations. *Ships and Offshore Structures* 12(5):697–705.

Chandrasekaran, S., Nagavinothini, R. 2017. Analysis and design of offshore triceratops under ultra-deep waters. *International Journal of Structural and Construction Engineering. World Academy of Science, Engineering and Technology* 11(11):1505–1513.

Chandrasekaran, S., Nagavinothini, R. 2018. Dynamic analyses and preliminary design of offshore triceratops in ultra-deep waters. *International Journal of Innovative Infrastructure Solutions.* DOI: 10.1007/s41062-017-0124-1.

Chandrasekaran, S., Nassery, J. 2017. Nonlinear response of stiffened triceratops under impact and non-impact waves. *International Journal of Ocean Systems Engineering* 7(3): 179–193.

Chandrasekaran, S., Nunzinate, L., Seriino, G., Caranannate, F. 2009. *Seismic Design Aids for Nonlinear Analysis of Reinforced Concrete Structures*. Boca Raton, FL: CRC Press. ISBN: 978-14-398091-4-3.

Chandrasekaran, S., Srivastava, G. 2017. *Design Aids for Offshore Structures under Special Environmental Loads Including Fire Resistance*. Singapore: Springer. ISBN: 978-981-10-7608-4.

Chapman, S.J. 2015. *MATLAB Programming for Engineers*. Boston, MA: Cengage Learning.

Derucher, K., Kim, U., Putcha, C. 2013. *Indeterminate Structural Analysis*. Lewiston, NY: The Edwin Mellen Press.

Fox, L. 1969. Computer Solution of Linear Algebraic Systems. By G. Forsythe and C.B. Moler, pp. xi, 148. 1967. Prentice-Hall. *The Mathematical Gazette* 53(384): 221–222.

Ghali, A., Neville, A., Brown, T. G. 2014. *Structural Analysis: A Unified Classical and Matrix Approach*. Boca Raton, FL: CRC Press.

Hibbeler, R.C., Kiang, T. 1984. *Structural Analysis* (Vol. 9). Upper Saddle River, NJ: Prentice Hall.

Holzer, S.M. 1985. *Computer Analysis of Structures: Matrix Structural Analysis Structured Programming* (Vol. 21). New York: Elsevier.

Householder, A.S. 2013. *The Theory of Matrices in Numerical Analysis*. Dover Publications.

Jarquio, R.V. 2007. *Structural Analysis: The Analytical Method*. Boca Raton, FL: CRC Press.

Karnovsky, I.A., Lebed, O. 2010. *Advanced Methods of Structural Analysis*. Berlin: Springer Science & Business Media.

Kassimali, A. 2009. *Structural Analysis*. Boston, MA: Cengage Learning.

Leet, K., Uang, C. M., Gilbert, A.M. 2002. *Fundamentals of Structural Analysis*. Chichester, UK: McGraw-Hill.

Livesley, R.K. 2013. *Matrix Methods of Structural Analysis: Pergamon International Library of Science, Technology, Engineering and Social Studies*. New York: Elsevier.

Lopez, C. 2014. *MATLAB Programming for Numerical Analysis*. New York: Apress.

Mau, S.T. 2012. *Introduction to Structural Analysis: Displacement and Force Methods*. Boca Raton, FL: CRC Press.

McCormac, J. C. 2007. *Structural Analysis: Using Classical and Matrix Methods*. Hoboken, NJ: Wiley.

McGuire, W., Gallagher, R.H., Ziemian, R.D. 2000. *Matrix Structural Analysis*. Hoboken, NJ: Wiley.

McKenzie, W.M. 2006. *Examples in Structural Analysis*. Boca Raton, FL: CRC Press.

Menon, D. 2008. *Structural Analysis*. Oxford: Alpha Science International.

Przemieniecki, J.S. 1985. *Theory of Matrix Structural Analysis*. Dover Publications.

Bibliography

Reddy, C.S. 2011. *Basic Structural Analysis*. New Delhi: Tata McGraw-Hill Education.
Sack, R.L. 1994. *Matrix Structural Analysis*. Long Grove, IL: Waveland Press.
Siauw, T., Bayen, A.M. 2014. *An Introduction to MATLAB Programming and Numerical Methods*. New York: Elsevier.
Smith, D.M. 2010. *Engineering Computation with MATLAB*. Boston, MA: Addison/Wesley.
Wang, C.K. 1962. *Statically Indeterminate Structures* (No. 531.12 W3). McGraw-Hill
Weaver, W., Gere, J.M. 2012. *Matrix Analysis Framed Structures*. Berlin: Springer Science & Business Media.
Wilson, E.L. 2002. *Three-Dimensional Static and Dynamic Analysis of Structures*. Berkeley, CA: Computers and Structures, Inc.

Index

A

Analysis of special members
   example problem, 210–219
   non-prismatic members, 209
      substructure technique, 209
   special elements, 207–209
      moment of inertia, 208
      Young's modulus, 208
   three-dimensional analysis of truss structures, 205–207
Axial deformation, 71

C

Continuous beam, 3
Cross-partitioning, 10–12

E

End moments and end shear, 79–80

F

FEM, *see* fixed end moments (FEM)
Fixed end moments (FEM), 78–79
Flexibility approach, 2
Flexural rigidity, 22–27

G

Global stiffness matrix, 76–77

K

Kinematic indeterminacy, 2–5
   stiffness approach, 2
      continuous beam, 3
      fixed beam, 3–4
      frame, 4–5
      simply supported beam, 4

L

Linear equations, 5–8
Local axes, 77

M

Matrix operations, 8–12
   banded, 12
   cross-partitioning, 10–12
   partitioning, 10
   submatrix, 8–10
Moment of inertia, 208

N

Non-prismatic members, 209
   substructure technique, 209

P

Planar non-orthogonal structures
   analysis, 77–80
      degrees-of-freedom, 78
      end moments and end shear, 79–80
      fixed end moments (FEM), 78–79
      local axes, 77
      reference axes, 77
      stiffness matrix, 78
      transformation matrix
         coefficients, 78
   example problems, 80–121
   global stiffness matrix, 76–77
   overview, 67–69
   stiffness matrix formulation, 69–71
      axial deformation, 71
   transformation matrix, 71–73
      end moments, for, 73–75
Planar orthogonal structures
   example problems, 30–67
      continuous beam, 30–38, 38–42
      orthogonal frame, 42–50, 50–54
      step frame, 55–60, 61–66
   flexural rigidity, varying, 22–27
      rotational coefficients, 27
      stiffness matrix, 22, 27–30
   indeterminacy, 1–5
      kinematic, 2–5
      static, 2
   linear equations, 5–8
      inverse of matrix, 6–7
      solution, 7–8
   matrix operations, 8–12
      banded, 12
      cross-partitioning, 10–12
      partitioning, 10
      submatrix, 8–10
   overview, 1
   rotational coefficients, 19–22

standard beam element, 12–19
    static degree-of-freedom, 14
Planar truss system
    example problems, 125–157
    overview, 123
    stiffness matrix, 124
    transformation matrix, 123–124

R

Rotational coefficients, 19–22

S

Standard beam element, 12–19
Static indeterminacy, 2
    flexibility approach, 2
Stiffness approach
    continuous beam, 3
    fixed beam, 3–4
    frame, 4–5
    simply supported beam, 4
Stiffness matrix, 22, 27–30
Substructure technique, 209

T

Three-dimensional analysis of space frames
    Ψ angle, 173–174
    analysis, 174–177
        Ψ angle, 174
        direction cosines, 174
    beam element, 159–160
    example problems, 178–204
    member rotation matrix, 166–168
    overview, 159
        right-hand thumb rule, 159
    stiffness matrix, 160–164
    transformation matrix, 164–166
        direction cosines, 165
    Y-Z-X transformation, 168–171
        rotation matrix, 171
    Z-Y-X transformation, 172–173
Three-dimensional analysis of truss structures, 205–207
Transformation matrix coefficients, 78

Y

Young's modulus, 208